Lead Forward

Mobility Air Force Command Nodes for Complex Operations

MICHAEL J. LOSTUMBO, JEFFREY S. BROWN, STEPHEN W. OLIVER, JR.

Prepared for the Department of the Air Force
Approved for public release; distribution unlimited

For more information on this publication, visit **www.rand.org/t/RRA1677-1**.

About RAND

Research Integrity

Published by the RAND Corporation, Santa Monica, Calif.

Library of Congress Cataloging-in-Publication Data is available for this publication.

ISBN: 978-1-9774-1071-9

Cover: Senior Airman Kristof J. Rixmann/U.S. Air Force.

Limited Print and Electronic Distribution Rights

About This Report

The future command and control (C2) structure of the United States Air Force (USAF) forces must be designed to withstand and cope with attacks on U.S. capabilities and to effectively adapt to a rapidly changing battle space. It must address the physical and cyber vulnerabilities of command nodes and networks. And it must be designed to retain effective C2 as both the supply of available forces and the mission prioritization changes. These issues challenge current command constructs even with all nodes and links fully functional.

Air Mobility Command (AMC) is considering how to structure, train, and present its forces to prepare to operate in environments that are more complex than in previous conflicts. In a contested environment, AMC needs to consider the survivability and effectiveness of its own forces, while also anticipating its links with the forces it will support (fighter and bomber aircraft, as well as joint and coalition forces). In this report we address how AMC could structure and prepare command elements to operate effectively while still meeting the needs of all they support.

The research reported here was commissioned by AMC and conducted within the Force Modernization and Employment Program of RAND Project AIR FORCE as part of a fiscal year 2021 project, "Agile Lead Wing Configuration Options for Air Mobility."

This document should be of interest to those interested in the organization of USAF wings and the challenges of operating in a future contested environment.

RAND Project AIR FORCE

RAND Project AIR FORCE (PAF), a division of the RAND Corporation, is the Department of the Air Force's (DAF's) federally funded research and development center for studies and analyses, supporting both the United States Air Force and the United States Space Force. PAF provides the DAF with independent analyses of policy alternatives affecting the development, employment, combat readiness, and support of current and future air, space, and cyber forces. Research is conducted in four programs: Strategy and Doctrine; Force Modernization and Employment; Resource Management; and Workforce, Development, and Health. The research reported here was prepared under contract FA7014-16-D-1000.

Additional information about PAF is available on our website:
www.rand.org/paf/

This report documents work originally shared with the DAF on April 8, 2022. The draft report, dated on June 13, 2022, was reviewed by formal peer reviewers and DAF subject-matter experts.

Acknowledgments

The authors gratefully acknowledge the support and insights provided by Maj Gen Laura L. Lenderman. Throughout the project, we worked closely with our project monitor, Col Jeffery T. Menasco. We appreciate his advice and support. We benefited from many conversations and visits to AMC wings and other organizations, and we are deeply grateful to all who shared with us their expertise and time. An earlier version of this report benefited from careful reviews by Kristin Lynch and Lt Gen (ret) Mark O. Schissler.

Finally, we would like to thank the AMC Commander, General Mike Minihan, for the valuable time and perspectives he provided the RAND team. A command-wide leadership focus on overcoming real and difficult warfighting challenges was evident to the authors throughout this project.

Summary

Issue

United States Air Force (USAF) forces operating in forward areas must have command and control (C2) capabilities designed to withstand and cope with attacks on U.S. capabilities and to effectively adapt to a rapidly changing battle space. They must address the physical and cyber vulnerabilities of command nodes and networks. And they must be designed to retain effective C2 as both the supply of available forces and the mission prioritization changes. These issues challenge current command constructs even with all nodes and links fully functional. The USAF is developing a new force generation model to present defined forces, each of which will cycle through four stages of readiness over a two-year period that culminates in trained and ready units available for joint tasking. Defined forces will include C2 force elements (FEs), which are envisioned to be trained in preparation for the demands of future high-end operations.

Air Mobility Command (AMC), the USAF's lead command for the Mobility Air Force (MAF), is considering how to structure, train, and present its forces to prepare to operate in environments that are more complex than in previous conflicts. In a contested environment, MAF C2 structures need to consider the survivability and effectiveness of their own forces, while also anticipating how the MAF will support operations of the combat air force (CAF), primarily the fighter and bomber aircraft of the USAF, as well as the joint and coalition forces that also rely on MAF capabilities. A further complication is that some MAF forces will remain under AMC's operational control, and some MAF forces will be placed under regional commands. AMC asked RAND to

- identify the demands the new USAF agility concepts will place on expeditionary wings
- develop alternative wing C2 constructs for expeditionary MAF forces, some potentially disaggregated, in forward areas under threat of missile attack
- provide a qualitative assessment of the alternative constructs.

Approach

In this report we characterize the changing demands driving AMC redesign of C2 elements. Taking into account both the USAF operating concept and the adversary challenges, we analyzed how future commanders' responsibilities and challenges might differ from those that were faced in recent conflicts in Iraq and Afghanistan. The report maps the implications of these changing demands on C2 FEs to inform the design and training of these elements. We then develop options to scope future MAF C2 capabilities. These options are assessed qualitatively, and the advantages and disadvantages of each option are discussed in the report and summarized in the box "Key Findings and Recommendations."

Key Findings and Recommendations

Findings	Recommendations
The MAF can add to the resilience of theater postures if it can increase the level of dispersion of the force and take advantage of civilian airfields and other locations not available to the CAF.	The MAF should develop a wing-level capability for C2 of independent MAF forces operating from multiple locations to prepare for future high-end fights.
A modular approach for MAF C2 FEs provides the most flexibility, given the wide scope of the global operations that MAF forces must perform.	The MAF should develop a modular approach to C2 FE.
Airpower should be centrally controlled at the highest level feasible. MAF units will be required to operate in situations where they cannot communicate with higher headquarters. This creates a demand for C2 FEs to be prepared to assume limited functions of higher headquarters under conditions-based authorities (CBAs).	• MAF C2 FEs should be prepared to assume limited operational planning functions of higher headquarters temporarily under CBAs. • MAF C2 FEs should be organized, trained, and equipped to reflect new demands on commanders in areas such as operational planning, logistics, operational deception, and recovery. • Communications will be contested and are a fundamental part of C2, so the MAF should consider whether the communications capabilities of the command element are suited to future operating environments.
The potential for attacks on AMC Global Air Mobility Support System (GAMSS) nodes forces new consideration of the GAMSS C2 footprint and authorities required for those nodes. They need to be able to redirect flights transiting GAMSS locations to respond to changing conditions and adversary attacks. In addition, GAMSS nodes need to be able to coordinate with intratheater airlift and joint logistics forces in order to link GAMSS-provided cargo with theater logistics capabilities.	AMC should work with the Joint Staff and regional commands to prepare for attacks on aerial ports of debarkation (APODs) and develop appropriate authorities and tactics, techniques, and procedures (TTPs) to allow the GAMSS nodes the flexibility to function in resilient ways and link with joint logistics.
Given the dynamic nature of future operations envisioned, MAF units and MAF aircrews will likely beddown at, employ from, or transit a wide range of adaptive basing options (see Chapter 2) led by commanders and C2 teams from multiple major commands (MAJCOMs). Therefore, it is critical that operating procedures be standardized USAF-wide for adaptive operations in order to maximize both force survivability and mission effectiveness.	AMC should work with other MAJCOMs to develop TTPs for C2 FEs operating in future complex environments. More broadly, the USAF should consider mechanisms to ensure integration of agility concepts and training across MAJCOMs.

Contents

About This Report ... iii
Summary ... v
Figures and Tables ... viii
Chapter 1. Command Challenges and Analytic Approach ... 1
- What RAND Was Asked to Do ... 3
- Approach ... 4
- Organization of This Report ... 4

Chapter 2. Future Operating Environments ... 5
- United States Air Force Command Principles ... 5
- Two Distinct Tasking Authorities ... 6
- Expeditionary Command in Recent Conflicts ... 9
- Agility Concepts for Future Conflicts ... 11
- Lead Wing Adaptation ... 13
- Mobility Air Force Operations in Future Conflicts ... 14
- Ways Agile Combat Employment Will Challenge Mission Command ... 17
- Adversary Challenges ... 19
- Implications for Command Elements ... 21
- Organizational Considerations of C2 FE Design ... 30

Chapter 3. Force Element Options ... 34
- Scope of Responsibilities ... 34
- How FE Scope Aligns with Mobility Air Force Priorities ... 36
- Summary of Advantages and Disadvantages of Different C2 FE Scopes ... 39
- Mobility Air Force Command at Spoke Locations ... 43
- Spoke Contribution to a Modular Force ... 45

Chapter 4. Findings and Recommendations ... 47
- Key Findings and Recommendations ... 47
- Next Steps ... 50

Abbreviations ... 53
Bibliography ... 56

Figures and Tables

Figures

Figure 2.1. Mobility Air Force Tasking Authorities 6
Figure 2.2. An Agile Combat Employment Cluster 11
Figure 2.3. Range of Mobility Air Force Operating Environments 16
Figure 3.1. Hub C2 Scope Options 35
Figure 3.2. Spoke C2 Scope Options 44

Tables

Table 2.1. Mobility Air Force Threats and Missions 17
Table 2.2. C2 Implications of Adversary Actions and Possible United States Air Force Adaptations 21
Table 2.3. Aligning Tasks and Organizations 31
Table 3.1. Qualitative Assessment of Options 40
Table 3.2. Hub C2 FE Scoping Options, Advantages and Disadvantages 43

Chapter 1. Command Challenges and Analytic Approach

The United States Air Force (USAF) has periodically faced the question of how to organize its forces so that they can quickly deploy to contingencies ready to operate. In this report we address one aspect of this problem: the design of wing-level command and control (C2) elements for Air Mobility Command (AMC) forces.

The current threat motivating USAF organizational change is the arsenal of long-range, precise missiles that Russia and China have been developing and that could be used to attack airfields and the forces they host. Such threats call into question the ability of the USAF to generate sufficient combat power to prevail in future conflicts. Several USAF organizations have published concepts designed to explain new strategies for operating in such an environment. These agile combat employment (ACE) concepts stress the importance of dispersal, movement, and recovery of airfields.[1] At the same time, the USAF has decided to move to a force generation model (FORGEN) that will group USAF forces into force elements (FEs), each of which will cycle through four bins (reset, prepare, ready, and available) that represent different stages of unit readiness.[2] AMC is considering design options for a wing-level C2 FE to employ in future ACE operations and, if pursued, how it should be designed.

Currently, most USAF garrison wings are organized along functional lines in four major groups: Operations, Maintenance, Mission Support, and Medical. Each wing is led by a wing commander, supported by a small command staff,[3] with four group commanders each overseeing one or more squadrons reporting to the wing commander. The Operations Group, in an operational wing (i.e., flying wing), includes intelligence, planning, and the aircrews who operate the wing's aircraft. The Maintenance Group is responsible for both aircraft and equipment maintenance. The Mission Support Group is the most diverse and is comprised of seven support functions that together maintain the base facilities and take care of the people who live there.[4] Finally, the Medical Group oversees the direct medical support to the wing.

---

[1] See Air Force Doctrine Note (AFDN) 1-21, *Agile Combat Employment.*

[2] These four bins represent a cycle for a unit that is meant to culminate in a fully manned, trained, and equipped unit ready for mission taskings in the final cycle.

[3] The wing staff organization includes offices devoted to public affairs, safety, history, legal, command post, chaplain, information protection, plans, equal opportunity, and inspector general (Air Force Instruction [AFI] 38-101, *Manpower and Organization*).

[4] The squadrons under the Mission Support Group are force support, civil engineer, communications, contracting squadron, logistics readiness, security forces, and aerial port (if applicable) (AFI 38-101, *Manpower and Organization*).

Forces from in-garrison wings in the United States can be sent abroad for contingency operations.[5] Typically entire wings are not deployed, but instead functional capabilities[6]—normally sub-squadron level—are drawn from multiple wings and assigned to create an air expeditionary wing (AEW). This is the method the USAF uses to organize C2, operational forces, and combat support and present them to a combatant commander through the relevant service component commander.[7] In other words, AMC has peacetime garrison wings that are designed to organize, train, and equip forces so that they are ready to be deployed to contingencies, where they are then re-formed into AEWs to conduct operations.[8]

The contingency AEW command structure resembles the peacetime wing structure, with a few notable differences. Peacetime wings tend to have one type of aircraft, while AEWs often have diverse aircraft types. AEWs tend to be tailored forces, in that the size and type of the forces match their mission, while the size of an in-garrison wing is more standardized.[9]

AEWs also include several operations centers devoted to specific tasks:

- Command post/communications focal point, a full-time communication node directly responsible to the commander, which serves as the focal point for orders, information, and requests relevant to the C2 of forces[10]
- Current Operations Mission planning cell, which consists of
  - wing-level planners supporting the wing's squadrons
  - mission planning
  - intelligence for debriefing and reporting.
- Maintenance Operations Center, which tracks and coordinates AEW-assigned aircraft maintenance and plans use of shared resources
- Reception Control Center for new forces joining the wing, or for USAF and joint forces that are coming into the theater and will continue on to a forward location
- Cargo Reception, which tracks inbound supplies
- Base Defense Operations Center, which oversees installation security activities.

Over the last two decades, the USAF has relied on AEWs to organize forces operating in support of the conflicts in Afghanistan and Iraq. For those conflicts, USAF forces tended to

---

[5] This has been USAF strategy since the late 1950s, but has been made more challenging since the end of the Cold War because the number of forward forces has been reduced, as has the overall size of the USAF.

[6] For example, the aircraft and maintenance might come from one unit, but support capabilities, such as logistics, engineering, or communications, might be drawn from other units or globally sourced from throughout the USAF.

[7] Department of the Air Force Instruction (DAFI) 10-401, *Air Force Operations Planning and Execution*, p. 29.

[8] An AEW consists of seven modular elements: airfield seizure, open the air base, C2, establish the air base, generate the mission, operate the air base, and robust the air base. DAFI 10-401, *Air Force Operations Planning and Execution*, p. 29.

[9] USAF aircraft squadrons are fairly standardized in terms of the number of aircraft and the aircraft type. Although USAF wing organization is standardized, the size of the wings is not. A wing could have from one to four aircraft squadrons.

[10] Air Force Manual (AFMAN) 10-207, *Command Posts*.

operate from a handful of major installations, which aggregated considerable air power in a few locations. Some of the benefits of this AEW construct were that it allowed the USAF to operate effectively and efficiently, while rotating forces into these war zones and then back to the United States. Garrison wings in the United States could continue to train and prepare forces for rotations with AEWs, while maintaining robust home-station operations and installation support functions. The USAF could tailor the forces for each AEW depending on the mission, rather than send a generic force package that might be misaligned with the mission or operational environment.

One perceived disadvantage of the AEW system used in Afghanistan and Iraq is that by putting together an ad hoc wing team, the assigned personnel and leadership had no opportunity to train together in advance of deployment to a war zone. While this was a particular problem as a new AEW formed, it also meant that over time the unit needed to train the new arrivals as the wing staff and operations centers, as personnel rotated in and out of the assigned units. In many instances, this constituted on-the-job training. Its cost was degraded expertise and efficiency during the time that the new team members learned their responsibilities. However, one benefit was that the organization could pass on to arriving personnel their insights to retain organizational advancements and specialized knowledge relevant to the mission, adversary, and local conditions, which might otherwise be difficult to transmit if whole units, rather than individuals, were rotated in and out of the theater.

In addition to correcting any deficiencies of the past, AMC is motivated to prepare for anticipated differences in the future operating environment. A future conflict with Russia or China is likely to involve precision missile attacks on airfields, cyberattacks, degradations to communications systems, and attacks on command nodes. As the potential for these attacks pushes the USAF to develop new operational concepts, AMC recognizes that it must anticipate how these adversary actions and USAF adaptation might stress future Mobility Air Force (MAF) command elements so that AMC can organize, train, and equip those elements to meet the demands of this new operating environment.

What RAND Was Asked to Do

Motivated by this combination of adversary actions and USAF adaptation, which could stress legacy C2 organizations and practices, AMC asked RAND to

- identify the demands the new USAF agility concepts will place on expeditionary wings[11]
- develop alternative wing C2 constructs for expeditionary MAF forces, some potentially disaggregated, in forward areas under threat of missile attack
- provide a qualitative assessment of the alternative constructs.

---

[11] Due to the concerns about the threats to air bases in the future, the USAF is developing new agility concepts in order to make forces both more difficult for adversaries to target and more resilient to enemy attack. This will be covered in Chapter 2.

Approach

We reviewed developing USAF doctrine and concepts relevant to future conflicts to understand the demands likely to be placed on MAF forces in the future. We also reviewed literature on future operations and the challenges that countries such as Russia and China could pose in future conflicts. In addition, we also conducted over 50 interviews and group discussions with many people in both AMC headquarters and wings involved in experimentation relevant to future operations to get their views regarding future needs and current shortfalls in MAF capabilities.[12]

Taking into account both the USAF operating concept and the adversary challenges, we analyzed how future commanders' responsibilities and challenges might differ from those that were faced in recent conflicts in Iraq and Afghanistan, and identified areas for attention for the training and the sizing of the MAF command FE accordingly, due to their new demands.

Next, we developed different scoping options for the C2 FEs, and then we developed criteria and evaluated the advantages and disadvantages of each option.

Organization of This Report

In Chapter 2, we discuss the future operating environment. We explain emerging USAF concepts for operating in this future environment, and we discuss the implications for future MAF operations. We also identify how the demands on future commanders will differ from current experience. In Chapter 3, we describe and assess the C2 FE scoping options, as well as the advantages and disadvantages of each option. In Chapter 4, we summarize our findings and recommendations, and discuss next steps.

---

[12] These included meetings at AMC headquarters, 18th Air Force, 618th Air Operations Center (AOC), 92nd Air Refueling Wing (ARW), 62nd Airlift Wing (AW), 19th AW, 22nd ARW, 374th AW, 86th AW, 603rd AOC, 22nd ARW, 317th AW, 34th Combat Training Squadron (CTS). Semistructured interviews covered consistent topic areas relevant to wing organizational functions and design, as well as wing adaptation to different threat situations. Each interview sought to elicit information from the areas of expertise of the participants.

Chapter 2. Future Operating Environments

This chapter describes evolving USAF concepts of command organization, as well as potential adversary attacks, in order to understand the additive demands that future operations might levy on wing-level command elements and below.

United States Air Force Command Principles

Long-standing USAF doctrine, practice, and operational experience have shown that airpower is most effective when centrally controlled. The advent of advanced computer and communication equipment allowed the USAF to achieve this ideal in recent conflicts. Operation Iraqi Freedom, which began in 2003, demonstrated the benefit of a common operating picture (COP).[13] A COP allows commanders to "answer three key questions with high confidence: Where are we? Where are our subordinates? And where is the enemy?"[14] These advances in communication and computer technology also came with organizational changes. For instance, before 1992, air mobility operations were controlled by geographically aligned organizations separating the control of forces in the United States and those in theater organizations outside the United States. Since 1992, global C2 of most air mobility operations has been consolidated in the 618th Air Operations Center (AOC)—initially designated the Tanker Airlift Control Center—located at Scott Air Force Base (AFB).

Centralization can be very effective in an undegraded environment, when the central command organization has the personnel, communications equipment, and expertise to plan, oversee, and assess operations. However, there are certain situations in which centralization can struggle and possibly become suboptimal. For example, when the command node is understaffed, it might become saturated and unable to process and act on the volume of information, which could reduce its effectiveness. A centralized system must guard against adversary attacks on central nodes and communications capabilities, which could quickly turn a well-functioning command system to a highly fractured one. That is, the peripheral nodes are structured to interact with the center, but not one another, so in the absence of the central node, they will struggle to adapt.

The USAF has several forward AOCs to control theater air operations. Three of the USAF's key AOCs are located close enough to potential adversaries that they are within range of weapons that could be launched from adversary territory. The 603rd AOC at Ramstein Air Base is located within range of Russian conventional missiles. Likewise, the 609th AOC sits within range of Iranian missiles, and the 607th AOC sits within range of North Korean long-range artillery capabilities. Other AOCs, such as the 613th AOC in Hawaii, could be vulnerable to

---

[13] Ackerman, "Operation Enduring Freedom Redefines Warfare"; Myers, "A Word from the Chairman."

[14] Pyles et al., *A Common Operating Picture for Air Force Materiel Sustainment*, p. 6.

ship-based attacks. However, AOCs anywhere in the world could be subject to terrorist, special forces, cyber, and/or communications-jamming attacks. MAF wings have an additional challenge in that they could operate under the C2 of different AOCs, as described in the following section.

Two Distinct Tasking Authorities

The USAF ideal is to achieve unity of command; however, MAF units could receive mission tasking originating from different command organizations at different times. In many instances, MAF units operate under the command of the U.S. Transportation Command (USTRANSCOM), in support of global mobility operations. In addition, assigned or deployed MAF forces could operate under the command of a theater or regional joint force commander (JFC). In view of a contingency, the JFC will typically appoint a joint force air component commander (JFACC) to be in charge of air operations in the JFC's geographic area of responsibility.[15] The JFC can also create organizations to oversee and coordinate joint logistics, such as a Joint Deployment and Distribution Operations Center (JDDOC). While MAF forces can be within these two distinct chains of command (USTRANSCOM or JFC), they can receive tasking from three sources: USTRANSCOM's Global Operations Center (GOC), through the 618th AOC; the JFC's regional JDDOC, through the JFACC's AOC/Air Mobility Division (AMD); or the regional JFACC, directly through the regional AOC/AMD. Figure 2.1 illustrates the two chains of command and some of the key organizations and relationships relevant to mobility operations.

Figure 2.1. Mobility Air Force Tasking Authorities

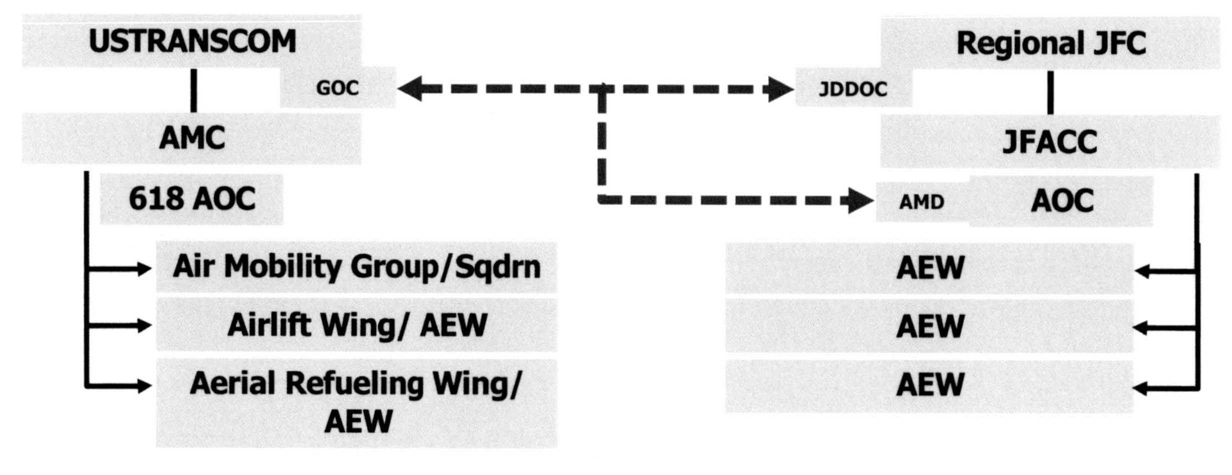

NOTE: JFC = Joint Force Commander; JDDOC = Joint Deployment and Distributions Operations Center; JFACC = Joint Force Air Component Commander; AOC = Air Operations Center; AMD = Air Mobility Division; AEW = Air Expeditionary Wing; GOC = Global Operations Center.

[15] Both the JDDOC and the JFACC align under the regional JFC, so are arguably within the same chain of command. We separate them here because they have related but distinct responsibilities. They coordinate their efforts, but may generate separate mission taskings to MAF forces assigned to subordinate air expeditionary wings.

Both chains of command have their own focus and authorities (global versus regional) and overlapping functions. Some MAF forces work exclusively within one chain of command throughout a contingency, but other forces may receive mission taskings from two of these chains of command, which occurred in some situations in Afghanistan and Iraq.[16] *Command relationships* (COMRELs) is the term used by the Department of Defense (DoD) to describe not just the chain of command, but the often-complicated relationships between forces operating together. Forces can be assigned to commanders with differing levels of authorities and responsibilities for both the commander and the forces. Just as important is that organizations and forces are designated as either supported or supporting. These supporting/supported relationships can define roles within a single command or across multiple commands.

The AMC global C2 node for USTRANSCOM-tasked air mobility missions is the 618th AOC at Scott AFB in Illinois. Although most AMC airlift and air-refueling aircraft are based in the continental United States, these missions take them around the world. In this sense, MAF aircrews are accustomed to operating alone or in small groups, and in conducting missions that could take them to many different countries. During these missions, the 618th AOC maintains in-flight visibility to track aircraft movement, re-tasks missions if requirements change, and helps to troubleshoot in case of any difficulties or unexpected situations.

The JDDOC typically supports the regional JFC—who in many cases is the geographic combatant commander—in developing and overseeing deployment and distribution plans within a theater or areas of operations.[17] The JDDOC receives all the requests for movement and resupply from forces in the theater, including requests from partner nations for U.S. support; and it both validates and prioritizes those requests and coordinates air, land, and sea movements to fill them.[18] The validation process means there is some filter between the users and the units executing the airlift missions, because airlift units are not tasked directly by users. The validation process is important in situations in which the demands exceed the supply of materiel and aircraft. Validated theater airlift missions can then be tasked to MAF units through either the Air Mobility Division (AMD) of the regional AOC[19] or, if coordinated with USTRANSCOM, via the 618th AOC.

Finally, the regional JFC normally delegates C2 of joint theater air operations to a JFACC, supported by an AOC. The JFACC identifies the goals of the air campaign and tasks forces with missions to accomplish those goals. A daily air-tasking order (ATO), which is prepared by the AOC, assigns missions to units with very detailed information about the mission and its goals. The JFACC provides direction to the force to define the commander's intent. The JFACC is

---

[16] In one interview, an aircraft commander with experience operating in Iraq and Afghanistan described receiving mission taskings from two different commands in one day.

[17] Joint Publication (JP) 4-0, *Joint Logistics*, p. III-9.

[18] The JFC can also establish a Joint Movement Center, which usually oversees the movement within a subregion of a theater.

[19] Recently, U.S. Air Force in Europe (USAFE) reorganized and renamed its AMD as the A-34.

provided authorities by the JFC, which can be delegated to lower echelons under certain conditions; these are known as conditions-based authorities (CBAs).

Although the JFACC has responsibility for all air operations, in practice the theater AOC focuses on planning and executing combat operations, while an organization within the AOC, the AMD, is expected to plan and C2 theater air mobility operations. The current AOC structure, with the AMD as a distinct division inside the AOC, reflects the historic evolution of these two organizations. The AOC was formally declared a weapon system and established in 2002. Before that, from Vietnam through the first Gulf War, airlift operated under a dedicated command that was separate from the combat forces. In Vietnam the commander of the 834th Air Division filled this role. In the Gulf War, the Commander Airlift Forces (COMALF) performed this role and was dual-hatted, serving both the JFC and USTRANSCOM. In both cases these commanders were supported by a C2 center that was near to, but physically separate from, the C2 center overseeing combat operations.[20]

Today, having the AMD within the regional AOC facilitates the synchronization of AMD-generated missions with other missions generated by the AOC and tasked via the ATO. Doctrinally, the AMD is responsible for planning all four types of core mobility operations: airlift, air refueling, air mobility support, and aeromedical evacuation (AE). However, in recent practice, theater tanker missions are planned in the AOC combat plans division and controlled by the combat operations division (COD) to achieve synergy with their primary users (e.g., fighters and bombers).

The USAF has a unique C2 problem compared with other services, because it relies on disaggregated forces to meet its objectives. The USAF routinely combines capabilities originating from multiple airfields separated by large distances to work together to meet common objectives. In contrast, other services employ forces in much closer proximity. By drawing from the capabilities of a disaggregated force, the USAF can efficiently employ airpower over very broad geographic areas. This approach is made possible by high-bandwidth communication among these distributed forces and a command concept that centralizes operational planning and decisionmaking. In this context, the key challenges for the commander are to obtain and process sufficient information to understand the battlespace, formulate ideas to achieve objectives that match the military problem and the available capabilities, and develop missions to implement those ideas and convey the ideas and missions to units to carry them out. This has been the

---

[20] See Devereaux, "Theater Airlift Management and Control." In Vietnam, the 834th Airlift Division served as Military Assistance Command Vietnam's theater airlift command organization. It had an Airlift Control Center at Tan Son Nhut, which was "the source of command and direction for the tactical airlift forces" (p. 16). In the Gulf War, a COMALF was dual hatted and performed what today would be described as the Air Force Forces (AFFOR) and JFACC functions for five airlift wings in the theater (pp. 32–33).

essence of the command concept that the USAF has used in recent conflicts in Afghanistan and Iraq.[21]

Expeditionary Command in Recent Conflicts

The USAF was able to meet the airpower demands of recent conflicts by employing force from a handful of major installations in the theater, each organized as an AEW, supported by global airlift and, if necessary, global strike from long-range assets based outside the theater. These installations had large numbers of different types of aircraft. Although the AEWs featured co-located combat air force (CAF) and MAF forces, the input to the ATO that provides their taskings came from different parts of the AOC: Airlift forces came through the AMD, while the CAF and aerial refueling forces came through the COD. Currently, there is some consideration of revising this arrangement so that in the future, all MAF forces assigned to an AEW would be tethered under the command of the AEW commander.[22]

These consolidated AEWs also often had Army or other service personnel on the installation, and in some cases other services were designated the lead service and given responsibility for operating the installation. The primary threats to these installations were ground-based, so they used security forces to monitor areas around the installation to defend against rocket and mortar attacks and control access to the installation.

The use of AEWs perpetuated the long-standing USAF practice of co-locating forces with their command (usually wing) element at an installation. When they are separated, it is usually considered temporary because such detachments are generally limited to aircrews and maintenance personnel and thus lack other key capabilities and expertise required to operate an installation. As we will discuss in the following section, ACE may break this tight link between location and command.

The USAF commander at an AEW has many responsibilities. These include ensuring that the installation remains capable of supporting air operations, that assigned aircrews and aircraft

---

[21] This highly centralized command concept has evolved as the technology has permitted. The USAF has a long history of adaptation in command. During World War II, before it was a separate service, the Army Air Corps tactical command structure changed over time and adapted to suit the theater of operations. In the European theater, the close distances and heavy adversary resistance favored larger air packages. As a result, the command structure adapted to coordinating large packages, which sometimes involved hundreds of aircraft. In contrast, very different employment and organizational structures were developed for the Pacific, with its vast distance, small airfields, and the limitations of Japanese air power. For this "Jungle Air Force" forward forces operated in independent groups. The Pacific air campaign "was characterized by (1) wide dispersion of units, (2) squadrons and groups operating far forward of their higher commands, (3) primitive and often temporary forward fields, (4) limited navigation aids and weather information, (5) long overwater missions, (6) severe tropical weather, and (7) a wide range of missions (including maritime interdiction)" (Vick, *Force Presentation in U.S. Air Force History and Airpower Narratives*, pp. 16–21).

[22] 317th AW, "After Action Report for Operation RAPID FORGE 2019." Air Combat Command (ACC), "Lead Wing Concept of Operations," v2.0, p. 22, implies this. See also 39th Airlift Squadron, "After Action Report for MOSAIC TIGER."

execute their ATO missions, and that higher headquarters has accurate current force readiness and postmission reporting so that it can assess operational effectiveness and plan future operations.

A USAF commander can also choose to create an air expeditionary task force (AETF), which is a layer of command between the AEW and the higher headquarters they both report to. An AETF would normally have responsibility for multiple AEWs. An AETF is an optional layer of command, but it could be particularly useful in a very large theater in which the geography of the theater segments the force.

Most USAF missions for Iraq and Afghanistan were planned at higher headquarters. For MAF units, these taskings could come either from the U.S. Central Command (USCENTCOM) chain of command through the 609th theater combined AOC, or from USTRANSCOM through the global 618th AOC. These commands developed processes and used technology to employ the force—that is, to design and allocate missions to meet the intent of their respective commanders. In either case, the tasking to the AEW would be similar. AEWs would be sent detailed mission taskings via the ATO for their forces to execute. In those cases and under current doctrine, wing command focuses on generating aircraft to meet scheduled missions and, by design, is divorced from force employment prioritization decisions made at the higher headquarters.[23]

Whether receiving mission tasking from USCENTCOM or USTRANSCOM, MAF units operating in support of operations in Iraq and Afghanistan experienced tight control of their actions. Aircrews experienced this tight control in several ways. They were not expected to plan their own missions. Rather, they were subject to "positive launch procedures," which required them to obtain explicit approval from the AOC at the time of mission launch and ask permission from the AOC and air traffic control to take off at each stop along a multistop mission. This is an example of the extent to which centralized control pervades even routine operating activities such as aircraft takeoff.[24] Similarly, when disruptions occurred, aircrews were not empowered to make adaptive decisions on their own, but instead were expected to provide updated information to higher headquarters (either to the 618th AOC or the theater AMD) and await their revised orders.

---

[23] The USAF distinguishes between force generation and force employment, which allows the command responsibility for these functions to be separated. Force generation refers to everything needed to be in a position to launch a sortie, while force employment involves choosing the mission for that sortie. Force generation is meant to ensure that "operational forces are properly organized, trained, equipped, and ready to respond to emerging crises and sustained operations" (DAFI 10-401, *Air Force Operations Planning and Execution*, p. 37). Force employment is "the strategic, operational, or tactical use of forces" (Department of Defense, *DoD Dictionary of Military and Associated Terms*).

[24] This positive launch procedure is codified in Air Mobility Command Instruction 11-208, *Mobility Air Force Management*, p. 48.

Agility Concepts for Future Conflicts

While the current USAF deployed command concept could face numerous challenges from adversary actions in future conflicts, it also needs to adapt to emerging changes in USAF operating concepts. Due to the concerns about the threats to air bases in the future, the USAF is developing new agility concepts in order to make forces both more difficult for adversaries to target and more resilient to enemy attack. ACE encompasses a number of interrelated agility concepts that numerous USAF MAJCOMs are developing. The first to be published came from the Pacific Air Forces (PACAF), but other ACE concept documents have been issued by the Headquarters Air Force and the U.S. Air Force in Europe (USAFE). Each one has unique features, but several common elements relate to wing command organization. These include an emphasis on dispersal, movement among airfields, and recovery.[25] The PACAF ACE document describes the need to prepare the theater for a future conflict and to prepare the force to operate in that conflict. One significant force-posturing change is the idea that forces should be postured in clusters. In this ACE approach, airfields are grouped in hub-and-spoke locations. The forces assigned to the hub and the spokes in these clusters would likely be organized under a single AEW. Figure 2.2 illustrates this concept. The main wing element will operate out of a central hub, but have delegated CBAs from its commander to move and disperse its assigned forces as needed among the hub, and several smaller spokes in order to execute air operations and counter adversary threats and activity.

Figure 2.2. An Agile Combat Employment Cluster

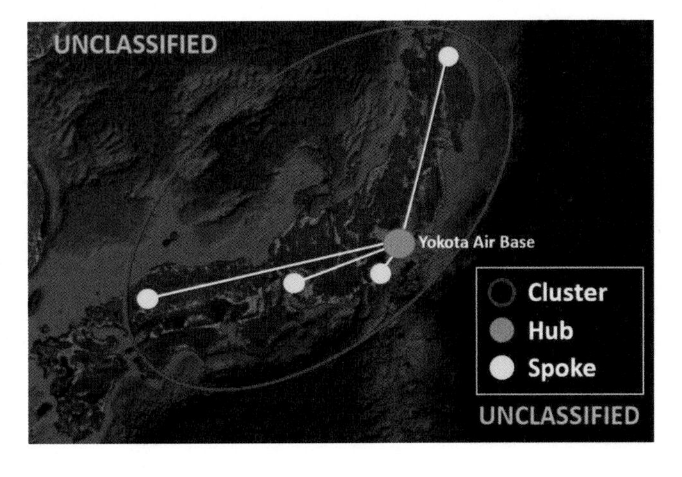

SOURCE: Reproduced from PACAF, *ACE*.

[25] See for instance, PACAF, *Agile Combat Employment (ACE): PACAF Annex to Department of the Air Force Adaptive Operations in Contested Environments*. (Hereafter PACAF, *ACE*.)

The purpose of ACE is to increase survivability of the force in the face of attacks, while maintaining the ability to generate combat power.[26] For the MAF, a hub-and-spoke beddown may sometimes make sense, but at other times it may not for two reasons. First, the density of MAF forces depends on the demands of the users globally, balanced by the threats. In many instances, the threat will push the MAF to smaller footprints at each location. For instance, if a CAF fighter wing is operating in a hub-and-spoke cluster, the MAF intertheater lift support to that cluster might be very modest, such that perhaps one MAF spoke would be sufficient. Second, the MAF is already accustomed to operating in small detachments for extended periods of time. Small detachments are more survivable than large ones, and if the MAF can generate combat power from such postures, it will meet the intent of ACE. Although we have focused on the ACE hub-and-spoke concept, in some situations it may make more sense to disperse forces to a few airfields, without necessarily needing a hub. This approach would be well suited to high threat areas because it would remove the hub as a potential single point of failure.

The ACE concept differs from recent expeditionary employment concepts in that the ACE AEW will be assigned responsibility for a cluster of airfields, breaking the practice of tying command to a single location. The AEW commander may well be in command of geographically separated forces, and conversely, detachments may operate for extended periods independently from their wing leadership control, employing CBAs, or mission-type orders (MTOs),[27] when normal communications links are lost, if such authorities have been predelegated to the detachments. The concept envisions hubs as the "conduit between forward operating locations (spokes) and rear echelon forces," which "must prepare to both receive follow-on forces and disperse current assets within a cluster's hub and spoke network."[28]

The agility concepts also emphasize the need to empower lower echelons to make decisions. This is a stark departure from the close control experience of recent conflicts described above. For instance, the PACAF document states, "The principle of commander's intent succeeds in PACAF when the commander generates and communicates intent well ahead of need and thrives in a steady-state culture of delegation down to the lowest component level."[29] This idea is more clearly stated in the relevant Air Force doctrine: "It is highly expected that elements conducting ACE will lose connectivity with operational C2; therefore it is imperative that units be trained to operate via commander's intent with limited direction from air operations centers or air component staff."[30]

---

[26] PACAF, *ACE*, p. 2.

[27] AMC has recently published a primer on MTOs. One circumstance for their employment is to "execute the assigned mission when conditions prevent mobility forces from communicating effectively and/or in a timely manner with higher level C2" (Minihan, "AMC Mission Type Orders Primer," p. 2). We see CBA as essentially synonymous with MTO, though the term "CBA" is not used in the AMC Primer.

[28] PACAF, *ACE*, p. 9.

[29] PACAF *ACE*, p. 8. Commander's intent is defined in JP 3-0, *Joint Operations*.

[30] AFDN 1-21, *Agile Combat Employment*, p. 7.

While we agree that the Air Force doctrine should highlight the need for units to operate with limited direction, we see the threat stemming not just from the possibility of lost "connectivity," but also due to the centralization of Air Force command and the vulnerability of its command centers to attack. The Air Force relies on high-bandwidth communications pathways in order to operate. These come primarily from fiberoptic cables, supplemented by satellite communication, and from a wide range of line-of-sight (LOS) tactical communications systems. These systems can certainly be attacked, either physically or through cyberattacks. Fiberoptic cables can be cut, and satellite communications radio receivers can be jammed, as can cellular communications. And forward units should be aware of how their communications emissions might be providing valuable information to the adversary. The USAF has several AOCs within range of missiles from adversary territory. Others are vulnerable to submarine missile attacks. And all could be attacked by special forces. Due to these threats, the USAF is considering how forces cut off from higher headquarters should respond. It remains a difficult question because centralized control is so fundamental to USAF operating concepts. While logically it is understood that the integrity of the centralized system cannot be guaranteed in the future, any adaptation would be a major change.

Lead Wing Adaptation

In response to the ACE concepts developed by PACAF and USAFE, Air Combat Command (ACC) has developed a Lead Wing Concept, and designated five wings to begin implementation.[31] These are intended to be "ACE-capable" trained and ready forces for support to a JFACC. "[Lead Wings are] organized, trained, and equipped to deploy from home station in accordance with Immediate Response Force (IRF) readiness table requirements, fall-in on an 'established airfield' (AFFOR responsibility), provide wing-level C2, agile force beddown, airbase defense and the ability to operate an airbase for up to 30-days in order to support rapid sortie generation of attached [forces]."[32] The concept provides considerable detail on how units will be organized to deploy and operate in a contested environment. These units will merge the operations and maintenance groups into mission-generation FEs, which are meant to be modular forces that can attach to C2 FEs and air base squadron (ABS) FEs.[33] The lead wing is sized to command a

---

[31] They are 4th Fighter Wing, Seymour Johnson AFB; 23rd Wing, Moody AFB; 55th Wing, Offutt AFB; 355th Wing, Davis-Monthan AFB; 366th Fighter Wing, Mountain Home AFB. ACC, "Air Combat Command Names Lead Wings."

[32] ACC, "Lead Wing Concept of Operations," v2.0, p. 4. Note that "IRF" is a DoD designation applied on a rotational basis to units that are prepared for immediate operational taskings by the Secretary of Defense.

[33] Force elements are the units that the USAF offers to the Joint Staff and Combatant Commanders for tasking. See DAFI 10-401, *Air Force Operations Planning and Execution*. C2 and ABS FEs are examples of types of FEs relevant to wing structure. By dividing the functions of C2, MG, and the air base squadron, the USAF maintains modular flexibility. For instance, two mission-generation FEs from different wings would be attached to a C2 FE.

squadron-size element of about 24 fighter aircraft. It can create a 12-ship detachment intended to operate for a short time at another location without organic ABS support.[34]

The 19th Airlift Wing (AW) participated in ACC exercises to test and develop the ACC Lead Wing Concept.[35] Working within its assigned table of organization and equipment, the 19th AW developed its own version of an MAF lead wing structure. Like the ACC version, it is designed to deploy to a theater hub (or main operating base) and operate in an ACE environment. It consists of a C2 FE and merges the operations and maintenance groups into a mission-generation (MG) FE and an ABS FE. Like the ACC model, the 19th AW concept envisions a C2 FE of about 50 people, with the wing commander being supported by staff organized under A-2, A-3/5, and A-1/4/6 directors, as well as most of the wing special staff.[36] In addition to exercising with CAF forces using an initial version of the concept in May 2021, the 19th AW held tabletop exercises in December 2021 and March 2022 to refine the concept.[37]

The 19th AW concept provides AMC with one option to consider in the design of its future capabilities. As noted, it was created using the existing personnel from a (relatively large) existing wing and so has the benefit of not increasing USAF force structure. Current AMC Air Mobility Operations Squadron unit type codes provide useful references to the design of the C2 FE regarding the needed subject-matter expertise, manning, and equipment to ensure the basic modular MAF C2 FE is capable of assuming limited AOC/AMD operational planning and coordination functions. As discussed in the following section, in future conflicts, the MAF will be performing all of its core functions from locations that will differ in many ways. It should consider a range of options for meeting future requirements. In considering those options, it should anticipate the operating environment and the ways that it differs from recent experience, which will inform not just the organization, but also the training that it will need.

Mobility Air Force Operations in Future Conflicts

While this report focuses on the demands of wing-level command and below, there are several different types of forces within AMC, and the command challenges for each deserves attention. In any future contingency involving Russia or China, AMC wings will be conducting four core functions: aerial refueling, airlift, air mobility support (the ground receipt of airlift), and AE.[38]

The USAF relies on tankers so that forces operating from long range can reach operating areas in the theater and/or extend their mission loiter time within a joint area of operations.

---

[34] ACC, "Lead Wing Concept of Operations," v2.0.

[35] Cohen, "New 'Lead Wing' Deployment Plan for Combat Aircraft Is Being Tested, Refined."

[36] 19th Airlift Wing, "MAF C2 Force Element Table Top Exercise (TTX)."

[37] The latter was observed by two of the authors.

[38] AFDP 3-36, *Air Mobility Operations*, p. 6.

Aerial refueling supports fighter and bomber operations, as well as airborne early warning, C2, and certain intelligence, surveillance, and reconnaissance and special operations aircraft. It is particularly relevant to operations in the Pacific theater, because the long distances and lack of basing infrastructure tend to emphasize disaggregated forces originating from geographically dispersed airfields operating from long range. Although the primary mission of tankers is aerial refueling, all USAF tanker aircraft have cargo holds, which can be used for airlift or AE. This inherent lift capacity enables tanker units to self-deploy.

Airlift operations execute cargo and passenger air movement. Airlift aircraft can move materiel either between theaters or within a theater. Early in a conflict airlift missions will support the deployment of forces and materiel into a theater. As the size of the deployed force grows, the airlift missions will increasingly support deliveries of supplies to sustain those forces. Airlift aircraft also conduct AE, a mission that could become increasingly important in a future war fight and could force future commanders to make very difficult decisions pitting the airlift requirements of evacuating seriously wounded Americans against the airlift needs of the ongoing war fight.

AE is beyond the scope of this report, although it is an AMC core function and could have implications for MAF C2 FEs. All air refueling and airlift aircraft can be used to support AE missions, although there are different procedures for moving critically wounded passengers. In some instances, the assigned missions may follow a pattern of inbound materiel airlift missions and outbound aeromedical evacuation.

Air mobility support for a contingency facilitates the movement of people and materiel from the United States to the theater of operations. It is handled through AMC's Global Air Mobility Support System (GAMSS), which includes a global network of fixed en-route aerial ports and assets to quickly transport materiel. Throughout this network, air mobility support units employ C2 elements, ground handling equipment, and personnel to support rapid unloading and loading of aircraft. They also can provide aircraft basic maintenance services for airlift aircraft. GAMSS also deploys tailorable, specially trained contingency response forces to rapidly open airfields, establish C2, and operate mobile aerial ports where no suitable fixed capabilities exist.

In a contingency in the Indo-Pacific theater, for example, the MAF will have FEs simultaneously conducting all four core missions. As Figure 2.3 illustrates, some MAF forces will execute intertheater flights from the United States to/through existing en-route bases and other temporary aerial ports in order to accomplish the inter- and intratheater movement missions. Other MAF FEs will forward deploy and establish expeditionary AEWs. Still other forces will implement ACE employment concepts and be postured as part of hub-and-spoke groupings of airfields. Units conducting aerial refueling, airlift, and air mobility support may have different C2 needs. By their nature, different missions and geographic locations will have different threat profiles within the theater. Exemplar missions/routes under 618th AOC control are depicted in blue, while those under some form of theater control are depicted in purple.

Figure 2.3. Range of Mobility Air Force Operating Environments

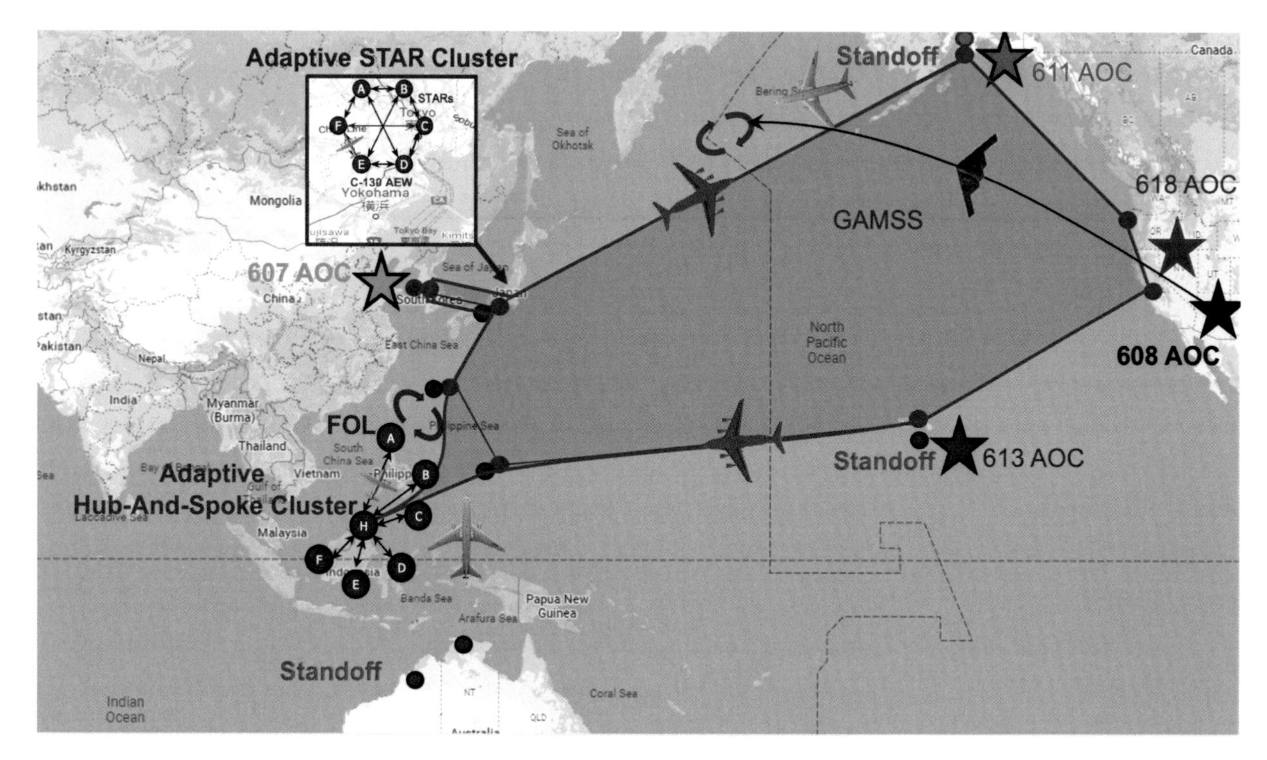

SOURCE: RAND analysis of discussions with AMC staff held November 9, 2021, and of briefing materials they provided.
NOTES: STAR = standard theater airlift route. Lettered purple circles indicate operating locations within each cluster, with location "H" representing the hub in a hub-and-spoke cluster. Two arrows forming a circle illustrate tanker orbits. Depicted locations of the continental United States AOCs are not geographically precise.

When operating in forward locations under threat of missile attack, the forces that are part of GAMSS will need the authorities and capabilities to disperse and shift locations and to be well integrated with joint theater logistics capabilities. These units act within AMC's chain of command, through the 618th AOC, but these new functions to implement ACE require tight integration with the theater C2 system under the JFC. AMC calls its concept for agility of current fixed en-route units Nodal ACE, which it defines as the ability to maneuver those forces to support joint power projection with resilient nodes. The potential for attacks on AMC GAMSS nodes forces a new consideration of the GAMSS C2 footprint and authorities for those nodes to be able to redirect flights transiting GAMSS locations as needed to respond to changing conditions and adversary attacks.[39] They also need new authorities to coordinate with intratheater airlift and joint logistics forces so that onward movement of GAMSS-provided cargo can be accomplished by theater logistics capabilities, when necessary.

Whether in an inter- or an intratheater role, airlift forces need a clear understanding of the chain of command and to be prepared to implement CBAs, which could include operationally allocating their forces. The task for airlift forces differs from that of GAMSS in that airlift forces

---

[39] The majority of AMC's global fixed en-route GAMSS nodes currently have little or no organic C2 capability. They rely on their USAFE/PACAF host wing's command post for both steady-state and contingency operations.

are focused on providing joint force users with unit movement/maneuver, sustainment, and installation supplies by air, either directly using intertheater airlift or through theater staging and intratheater delivery. The ability for airlift forces to implement delegated authorities is something that needs further development within the USAF.

Finally, the aerial refueling forces have a third task that differs from the other two. Aerial refueling focuses on providing aerial refueling support to CAF forces in the air while those forces are conducting their missions or transiting to or maneuvering within the theater. They are not focused on installations per se, the way that airlift forces need to be.

In addition to performing different functions in different chains of command, MAF forces will face a wide range of threat environments. The level of threat they face may change the way they are employed and the demand on their C2 FE. Table 2.1 below summarizes some of the dimensions of future MAF operations in a contested, degraded, and operationally limited environment. These differing demands and their implication for C2 FE design will be explored further in the following chapter.

Table 2.1. Mobility Air Force Threats and Missions

Location Type	Missile Threat	Communications Threat	En-Route System	Aerial Lift	Aerial Refueling
Standoff	Limited	Moderate	Link intertheater airlift with theater logistics	Support joint users and installations	Support CAF and joint missions
Hub	High	BLOS[a]: high LOS: moderate	Disperse and shift locations Redirect flights Link with intra-theater logistics	Support multiple users Prepare to operationally C2 forces	Support multiple users Prepare to operationally C2 forces
Spoke	High	BLOS: high LOS: moderate	N/A	Support multiple users Prepare to operationally C2 forces	Support multiple users Prepare to operationally C2 forces

[a]BLOS = beyond line of sight.

Ways Agile Combat Employment Will Challenge Mission Command

As noted above, MAF forces often operate as disaggregated forces. Single aircraft or small detachments of a few aircraft with planned support often deploy to and/or operate from distant military and/or commercial airfields, some of which may even be uncontrolled or unpaved. In some respects, this makes MAF forces well suited to ACE operations, but operating in dispersed postures could be more challenging in a future conflict with Russia or China for three reasons. First, MAF forces will not be operating in a benign environment. They can expect to be attacked, and the forces they support will also suffer attacks, so there will be a need to manage attrition of aircraft, equipment, and personnel. Second, the pace of operations and required decision speed will likely be dramatically increased. Third, their communications ability will also come under attack, and so connectivity to a centralized command, control, and mission planning/tasking node cannot be assumed.

The C2 elements designed for future conflict will need to account for geographic separation of the force under threat of dynamic physical attacks and cyberattacks. Command elements at hubs will need to have the capacity to monitor the airfields and forces within their cluster, allocate cluster resources among locations, communicate with higher headquarters and with users, and direct the movement of aircraft and capabilities. Spoke locations may also need enhanced C2 capabilities. These implications will be considered in Chapter 3.

If MAF wings operate from dispersed postures, the organizational layers of the wing increase. Instead of being co-located with all the forces of the wing together, a disaggregated force will either rely increasingly on communications technology or assume greater authority over its own operations, potentially including autonomous operations based on commander's intent. In addition, the organizational layers of the wing will increase, as each operating location will have some duplication of organizations and responsibilities of the wing hub.

Sustaining a disaggregated force becomes more complicated. Some items that are large and/or rarely used will not be readily available at locations with small footprints or locations intended for temporary use. And some detachments/forward operating locations may not need a full planeload of lift at any given time.

So-called lift-and-shift operations, or the moving of units from location to location, is another feature of agile combat employment.[40] This adds further complexity to those introduced by dispersal in that it could create a network of airfields and forces that are always in flux. While mobility operations make it difficult for the adversary to keep track of those forces, it also creates challenges for those trying to supply forces on the move. For instance, if a unit requests certain supplies, logistics planners not only need to source the request, but also need to be cognizant of any unit movements to ensure timely delivery to the right locations.

While the basic components of ACE are dispersal, maneuver, and recovery, the USAF could adopt a range of tactical practices in response to the different types of attacks described above. For instance, one adaptation to attacks focusing on aircraft on the ground is to develop capabilities to turn aircraft quickly and keep them in the air as much as possible. Such a tactic has considerable cascading demands. For the C2 elements directing continuous operations, some of the challenges include increasing the planning tempo to meet the increased operational tempo; more challenging demands to manage crew rest to be able to sustain operations over time; and increased demands for consumables and spare parts.

---

[40] AFDN 1-21, *Agile Combat Employment*, p. 3.

Another ACE-related adaptation involves the cross-training of airmen. In response to the need to both achieve greater levels of dispersal and reduce the footprint at locations in range of adversary missiles, the USAF has begun to cross-train airmen to accomplish additional MG or base operations support tasks normally performed by specialists from other career fields.[41] One of the early efforts has been to train nonfuel specialists to refuel aircraft, but that is just one example of an area under active exploration and experimentation throughout the USAF.[42] The C2 implication of using multicapable airmen for tasks is that the command element may need to account for and mitigate some downside risks. Tasks performed by multicapable airmen may take longer due to lack of experience or because they are also conducting their primary assigned duties. The work may also require more expert oversight to ensure that quality standards are maintained.[43]

Adversary Challenges

Part of the reason why preparing for future conflicts is difficult is that compared with recent conflicts, those future conflicts might involve a much wider range of possible attacks from peer adversaries. These include attacks on

- C2 nodes
- communications networks[44]
- computer systems
- aircraft
- airfield assets
- logistics hubs and stockpiles.

The concept of centralized command is the core tenet of USAF doctrine and operational employment.[45] It is so fundamental that it can be hard to imagine feasible alternatives. Thus, it is important to consider an adversary strategy to attack command nodes. In Europe, USAFE headquarters is within range of cruise missiles launched from inside Russian territory. PACAF

---

[41] So-called multicapable airmen, or "Airmen capable of accomplishing tasks outside of their core Air Force Specialty" (AFDN 1-21, *Agile Combat Employment*, p. 3).

[42] See, for instance, Underwood, "Putting ACE to the Test."

[43] Other adaptations may involve an ability to plan and implement camouflage, concealment, and deception (CC&D) to reduce the efficacy of adversary attacks. Included as part of this should be a capability to obscure the location of command elements and a consideration of the value of mobility for the command post.

[44] An electromagnetic pulse attack would be an extreme example of an attack resulting in widespread communications disruptions. If such an attack were the result of a nuclear detonation, that would certainly not be treated as routine.

[45] The long-standing formulation of this tenet was recently updated to "Airmen execute *mission command* through *centralized command, distributed control,* and *decentralized execution.*" The full implications of adding "distributed control" to this tenet will only emerge over time, though it may simply be a reflection of current practice. This report focuses on wing-level command, which is traditionally not part of the distributed control being addressed in this definition. Distributed control has long been enacted through battle management nodes. Theater commanders normally have direct tactical control of forces, essentially bypassing the wing-level echelon.

headquarters is not in range of conventional systems launched from within China, but it could be attacked by missiles launched from submarines. In addition, AOCs can be attacked by special forces or be subject to cyberattacks.

Communications pathways are also vulnerable, and adversaries are expected to target these. DoD has highlighted that authoritative military reports from China discuss "destroying, damaging, and interfering in the enemy's . . . communications satellites."[46] Most long-distance communications rely on fiberoptic cables, which provide some of the most responsive communications channels.[47] The locations of fiberoptic trunk lines are known and can be attacked.[48] Satellites provide an alternative means of communications, of which there are several different kinds, including wideband, which covers a large area on the ground; narrow band, from low earth-orbit satellites that provide coverage through frequent overflight; and advanced extremely high frequency, which is provided by large geosynchronous satellites.[49]

The C2 implications of these attacks for the USAF are far-reaching. These could include degraded or disrupted communications; fragmented chains of command, due either to attacks on command nodes or to network degradations; data integrity loss, due to cyberattack; and logistics scarcity, due to attacks on logistics nodes. In addition, the need to manage potentially complex airfield recovery operations may arise, as could an expanded planning burden, due to the need to develop more branch plans and to dynamically redirect current operations, given greater variability in aircraft availability and rapidly changing needs. Table 2.2 maps some possible adversary actions and USAF adaptations with specific C2 implications.

A final category of C2 challenges will come from adversary initiatives. In a conflict with Russia or China, unexpected operations could force the USAF to respond dynamically. This could mean having to jettison days of mission planning and could even lead to a reprioritization of the commander's requirements involving MAF forces and capabilities. In contrast with the experience in Afghanistan and Iraq, where many missions could be reliably planned in advance, unexpected operations on the part of Russia or China will require the USAF to make much more dynamic adjustments to current missions and to plan quickly for near-future missions.

---

[46] Office of the Secretary of Defense, *Annual Report to Congress*.

[47] 5th Combat Communications Group, *5 CCG Planners Guide*, p. 10.

[48] See Sutton, "How Russian Spy Submarines Can Interfere with Undersea Internet Cables."

[49] 5th Combat Communications Group, *5 CCG Planners Guide*.

Table 2.2. C2 Implications of Adversary Actions and Possible United States Air Force Adaptations

Adversary Actions	Possible Adaptation	C2 Implications
C2 node attack	CC&D Indications and warning (I&W)	Assume responsibility of higher headquarters Fragmented chains of command Increased mission planning Degraded COP
Communications network attack	Communications expertise and equipment for multiple alt-comms options, often in the form of primary, alternate, contingency, and emergency plans	Assume responsibility of higher headquarters Increased mission planning Ensuring logistics needs met more challenging Degraded COP
Cyberattack	Multiple diagnostic tools, defensive software, and user practices (often users are the first to identify anomalies)	Verification procedures if data integrity lost Denial of service could resemble consequences of network attack
Aircraft attack	Continuous operations Branch planning	Increase tempo of mission planning Ability to replace lost aircraft Dynamically re-task missions Develop and employ base CC&D strategy
Airfield attack	Recovery Branch planning I&W	Manage a complex recovery operation Dynamically re-task missions
Logistics node attack	Direct delivery Larger on-base stocks	Logistics not routine
Adversary initiative	Reactive dispersal	Multiday planning nullified

Implications for Command Elements

The C2 implications described in Table 2.2 can be grouped to identify some of the additional burdens that could fall on MAF C2 FEs as a result of such attacks. These include

- operational planning involving formulating and assigning missions to forces, especially if they are assuming responsibilities of higher headquarters
- mission planning, including the end-to-end details of each mission
- recovery, which means being able to manage complex response and recovery actions postattack
- logistics, or the ability to secure necessary resources in a communications-denied environment
- CC&D, which includes an ability to identify vulnerabilities of the installation to attack and to implement measures to reduce those vulnerabilities or their consequences

- intelligence, such as providing indications and warnings of attack, characterizing threats to an installation, and providing information relevant to missions
- communications, including an ability to operate equipment to enable communication through multiple channels
- command of disaggregated forces.

In the next section, we expand on the demands of each of these additional burdens due to the increased complexity of ACE operations, adversary attacks, and the possibility of delegated or conditions-based authorities.

Operational Planning

Activity

At the theater level, it is the commander's role to formulate strategy and assign missions to forces. Ideally this job is informed by a good understanding of the adversary forces and intentions and the capabilities within the command and the broader coalition. Operational planning involves deciding what is needed by whom and when, prioritizing the needs, and then matching assets to fulfill the needs. Currently, operational planning is done by command staff at higher headquarters, usually the AOC. Wings are assigned missions, which are often prescribed with a fair amount of detail. The wing assigns missions to ready crews, and the crews conduct mission planning.

Problem

The potential for lower echelons to assume some responsibilities of higher headquarters if they become disconnected from their chain of command has been anticipated in recent USAF policy and doctrine.[50] These documents envision this as a temporary problem and assume that wings will be able to reestablish contact with their higher headquarters or an appropriate alternate.

The discussion of MTO guidance tends to assume that commanders will impart a basic message to units that they should just keep doing what they have been doing, should they become disconnected. This might be feasible for some routinized missions, such as recurring missions along a fixed route, which are called standard theater airlift route (STAR) missions. (In peacetime, similar missions are called channel missions.) A unit participating in such STAR missions could continue flying those missions, but depending on the guidance the unit was, or was not, given in its MTO, the unit might, or might not, have a basis for prioritizing the actual cargo it carries on those STAR missions. A preference for unchanging missions is understandable; however, the possibility of missile attacks and cyberattacks on C2 nodes may not be easily or quickly resolved in all circumstances, so a short duration is not guaranteed. Furthermore, to

---

[50] CBAs are described in several recent USAF documents, including AFDN 1-21, *Agile Combat Employment*; PACAF's *ACE*; Minihan, "AMC Mission Type Orders Primer."

operate effectively in the dynamic battlefield expected in the future, the mission tasks will also need to change dynamically.

Implications

Operational planning at lower echelons places additional burdens on those units to gather information about their users' needs in order to establish an accurate demand signal to use as a basis to allocate missions. The planning staff would need to be equipped to gather this information from users. The unit may need to exploit other avenues for information. Returning aircrews will likely have some relevant tactical information. Similarly, airfields will periodically receive airlift missions. A carrier pigeon–type communication network is often discussed, but a system to efficiently exchange information along an airlift route has not been devised. To some extent the additional staff burden will be in trying to get information from alternative sources.

As described above, the validation of theater airlift requirements normally happens at the JDDOC, and then the AMD assigns airlift to service those needs. Under normal circumstances, the airlift community is told what to move and when to move it, but it is not necessarily told what it is for or why it is important. As a result, a wing commander has typically lacked the information necessary to validate user requests or to prioritize among competing users.

For an aerial refueling unit operating in a communications-denied environment, operational planning involves deciding where, when, and how many tankers to send. The receivers they support could be co-located or close enough to communicate by radio, in which case the coordination problem is solved, or be halfway around the world, as in the case of bombers. In the co-located case, the wing-level tanker planners would assume responsibilities that would normally be assumed in the AOC. For missions involving distant aircraft, and therefore a higher likelihood of being out of communication, the unit may have to send tankers on missions either deemed most critical or deemed most likely to succeed.

These new responsibilities will likely require a larger planning staff, which would scale more or less with the number of daily missions and users for which the wing is responsible. For instance, if the wing's squadrons became tethered to a user, the size of the wing's planning cell may not change all that much. However, if the wing's planning cell were to take on many or all of the functions of the AOC/AMD—especially in situations with multiple countries, multiple users, and multiple services, operating across a large area and in concert with other wings—then the wing's planning cell would necessarily grow significantly to be able to perform all or most of the functions that were previously handled by the AOC/AMD. This expanded role for the wings could require them to gain CAF integration experts, airspace planners, planners from other mission design series (MDS), host-nation liaisons, embassy liaisons, and logistics planners, to name a few. Operating this larger cell on a 24/7 basis would likely require a team of around 40–50 people (20–25 on each shift). It is conceivable that many of these people could be internally sourced within the wing, but they would need specialized training to operate optimally. Currently, the USAF trains people for these functions at the AOC course located at Hurlburt AFB. Potentially, this course could provide the baseline from which to create a new tailored course on how to operate a wing in coordination with other theater wings. And, of course, once trained (either formally or informally), the wings would need to be exercised to reinforce training and to hone procedures.

Mission Planning

Activity

The core mission-planning functions for mobility assets involve the planning of routes and times based on airspace, diplomatic clearance, weather, threats, aircraft capability, airfield capability, and so on. The loads are planned according to requirements and aircraft limits.

Problem

Mission-planning workloads will likely increase because of the need to develop more alternative options, due to the greater likelihood of forces experiencing friction of some sort when conducting missions. Potential differences from recent USAF experience are the pace of operations and lack of precise knowledge of requirements, status of airfields, and even aircraft, aircrews, and support personnel. The expectation that operations will be opposed kinetically and nonkinetically can dramatically change the efficiency and effectiveness of mobility operations. Maintaining operations in the face of ballistic/cruise missile and/or drone attacks and dealing with actual losses in the air are not events for which the USAF has grown accustomed. In the past, if an airlifter or tanker launched on a mission, it would accomplish that mission and return to base with near 100-percent certainty. In the future fight, that same aircraft may "disappear" inexplicably somewhere between airport A and airport B, having been shot down or destroyed on the ground at an en-route stop. Or it could have been diverted and be unable to communicate its location to others. A forward-based aerial port that up until a few moments ago was on the phone with a wing C2 element may go off the net with no explanation as to why. Were they attacked? Jammed? Overrun? Or just facing communications equipment problems?

Implications

These imagined situations and others create an environment that will place a premium on real-time threat-warning and intelligence to assist C2 in making better decisions about how to react and what to do in the face of such enemy action.

The linchpin is the ability to communicate with the users and with the providers. Ideally, there will be branch plans to anticipate needs and/or self-task in cases of degraded communications. In the tanker case, an MTO stating something like "Always keep 400,000 lbs. of gas airborne 24/7/365 over area A. For airlift, send 12 bombs, 45 tons of jet fuel, 250 MREs [meals ready to eat], and a pallet of spare parts to base B each day." For future airlift wings, developing a sense of diverse user needs and how to prioritize them could help in situations where planners struggle to keep mission taskings updated to meet rapidly changing needs.

Developing the capacity to handle emergency lift requests is likely to be an important skill for mission planners to have in future conflicts. It may well be that the demand for scarce lift and air-refueling assets, and in particular the demand for emergency lift, will be higher than it has been during the last 20-plus years. In Vietnam, when operations transitioned from routine to high-tempo, the lift demands could change dramatically, sometimes exceeding the capacity of the lift system to support operations. This had spiraling negative consequences, in that as users faced shortages, they responded by inflating lower-priority requests to "emergency" status, which undermined the prioritization process and increased the lift shortfall.[51]

Recovery

Activity

The implications of an attack and the demands of a major recovery operation are noted because they are likely to dominate the attention of wing leadership in the moment and call for leadership skills and resources that may be unique. In the USAF, Air Force Materiel Command (AFMC) has the lead for developing capabilities to recover after an attack. Although most MAF wings have organic elements that could make up a deployed ABS, a future lead wing could combine the MAF operations and maintenance groups with ABS capabilities drawn from another part of the USAF, such as AFMC. As a result, the ABS may not be viewed by AMC as part of the MAF C2 FE, but instead would be part of those responsible for the ABS element that, in most instances, would need to be joined with other force elements in order to have a fully functioning wing.

Every major USAF installation has a civil engineering squadron that is responsible for the physical infrastructure of the base, including all the built infrastructure (vertical and horizontal) and utilities (power, sewer, water, telecommunications infrastructure), as well as for expedient hardening (that is, hardening improvements accomplished after deployment), shelters, and CC&D.[52] In addition, the Air Force Civil Engineering Center has the lead in developing runway-repair capabilities. Engineers may also be involved in recovering from fuel storage attacks by building the facilities to operate fuel bladders or other expedient fuel storage. The ABS will also include the firefighting capabilities on the installation. Engineering and firefighting capabilities will be important for recovering an air base after an attack, but a major attack is likely to require the efforts of many outside the ABS.

Problem

The USAF has not experienced large-scale precision missile attacks on operating installations. Thus, although the ABS will have some of the capabilities necessary for a major recovery operation, it is not sized or trained to serve as the command center for a complex recovery event.

---

[51] Devereaux, "Theater Airlift Management and Control," p. 20.

[52] Air Force Pamphlet [AFPAM] 10-219, Vol. 2, *Civil Engineer Contingency Response and Recovery Preparations*.

This potential shortfall is noted because the expectation is that many air bases will fall under attack. Thus, assessing the implications for MAF wings is something that AMC should explore with other USAF major commands (MAJCOMs) in the future.

Implications

Recovery will not be a steady-state task for the C2 FE, but a major attack could require action by the entire installation to provide emergency response actions and to quickly recover the base. The USAF will need to continue to develop plans and capabilities to better position wings to recover from attacks. In conjunction, the MAF should consider the implication for the C2 FE design.

Logistics

Activity

Logistics involves a broad set of activities. One aspect of logistics is the means to order and receive needed supplies. In recent conflicts, the USAF could count on fairly routinized delivery of supplies. While land-locked Afghanistan did provide a difficult logistics challenge when materiel needed to be routed overland through Pakistan, for most of the conflict, USAF forces managed long-haul logistics. In many conflicts through its history, the USAF could expect logistics to be handled by routine actions. In future conflicts, the logistics system will need to adapt to a more dynamic situation. That is, it will need to move away from a model of routine pull-based logistics and be able to adapt to changing circumstances and demands. This could include so-called push-logistics, in which supplies are sent forward as circumstances change, without waiting to be told of new needs by forces that might be under attack or other stressful situations. For instance, a push-based system could prioritize outbound loads on a weekly basis via a JFACC cargo priority matrix. One week, the top priorities might be water; food; plasma; and small arms ammunition types a, b, and c, in that order. Secondary priorities might be medical supplies, jet fuel, diesel fuel, munition type x, munition type y, other required parts, servicing equipment, and/or expendables for friendly aircraft. The following week these priorities could, and would, be adjusted as necessary as determined by the JFACC.

Problem

The concern in future operations against Russia and China is that logistics will no longer be routine. Like recovery, logistics shortages could dominate wing leadership attention in periods of scarcity. Their own in-bound supplies could be delayed, and it could also pose challenges for mission planning of both airlift missions, which may arrive at a location to pick up supplies that have yet to be delivered, and tanker missions that will rely on fuel being available at a distant location. The readiness of MAF units depends in part on the timely delivery of supplies and spare parts in order to meet mission tasking. The wing leadership may need to track incoming supplies more closely and work with logistics readiness officers to find solutions to logistics disruptions.

Implications

In designing the MAF C2 FE, AMC should consider the extent to which the potential for logistics disruptions might hinder the readiness of MAF units and whether it warrants any adjustments to personnel or equipment to predict disruptions and find ways to circumvent them.

Camouflage, Concealment, and Deception

Activity

CC&D is another response to the vulnerability to missile attacks. It crosses the boundaries of existing wing organizational structure and could introduce command challenges that are not well appreciated in advance of a conflict. CC&D can cover a wide range of topics. The maintenance squadron could disperse aircraft around the airfield and under special coverings meant to make it more difficult for the adversary to accurately target aircraft. A special unit could install jammers at the air base to defeat active seekers on missiles. Or the engineering squadron could devise ways to conceal key capabilities or harden them against attack. All of these activities can fall under this broad concept of CC&D.

Problem

Currently, the USAF is not structured to provide someone with expertise to develop and implement a base-wide CC&D strategy.[53] At the wing level, the USAF does not have trained personnel able to implement CC&D actions broadly and develop an installation CC&D strategy.[54] Thus, some of the decisions to implement such a strategy will influence many aspects of installation operations. In many instances, CC&D tactics will increase the time to complete tasks (such as implementing on-base dispersal), which means that to be successful they will need to be enforced at a high level. The implication is that CC&D could be another area where planning and expertise that are currently lacking could suddenly become important to leaders in the field.

Implications

CC&D is a broad area that cuts across many different USAF specialties. To the extent that MAF wings will seek to implement CC&D measures, this lack of a natural home within current USAF structure might mean that the wing C2 FE could have to take on this role. AMC should

---

[53] The ACC LW concept envisions an Information Warfare Working Group, consisting of experts from within the wing, including cyberspace operations, electromagnetic spectrum operations, information operations, intelligence, weather, Office of Special Investigations, and others.

[54] Wings could, however, consult with USAF military deception specialists in support of developing a CC&D strategy.

consider how much CC&D planning and training need to be part of the C2 FE. Conceptually, this function could be placed within the OSS mission planning cell or embedded within the flying squadron, augmented by airfield operations, civil engineering, communications, and maintenance representatives with the rank and authority to quickly move, cover, and otherwise alter assets.

Intelligence

Activity

Wing-level intelligence support is provided by the combat intelligence center. This center is responsible for a variety of intelligence support, including intelligence preparation of the battlespace, mission reports, intelligence summary reports, situation briefings, and support to base defense. Squadrons also have intelligence support, with similar responsibilities tailored to the squadron needs.

Problem

The problem for intelligence officers at the wing and squadron levels reflects some of the same challenges facing the wings and detachments more broadly—namely, preparing for denied or disrupted communications and disaggregated operations. The intelligence unit is designed to draw from global intelligence capabilities. It relies on multiple sources of information to provide accurate assessments of adversary capabilities and intentions. Furthermore, it is expected to provide inputs in the form of mission reports so that others can draw insights from the experience of the operating units. A denial of communication abilities prevents intelligence professionals from accessing current information, although they can still provide valuable insights based on their experience and on accumulated information.

As a wing increases the number of operating locations, the wing and squadron intelligence complements will need to decide where to locate. Placing intelligence professionals at every location risks diluting their effectiveness, but keeping them centralized at one location risks communications outages with other locations.

In addition, although providing support to base defense is an intelligence function, wing- and squadron-level intelligence has traditionally focused on threats to aircraft in the air. As we emphasize in this report, the threats to airfields from a wide range of sources are expected to increase in the future. Thus, the intelligence support to base defense may require new capabilities and training to address these threats.

Implications

In supporting future forward-operating units, intelligence professionals will need to assess the implications of communications disruptions and disaggregated operations. Potentially, forward-deployed intelligence airmen could utilize the long-range communications capabilities onboard MAF aircraft equipped with tactical data links (such as dynamic re-tasking capability, real time information in the cockpit, and satellite communications) to reestablish secure communications. Regardless, the implications of a wider range of threats to airfields will need to be considered when designing the training and assessment capabilities for those units.

Communications

Activity

It is widely recognized that communications capabilities are critical to USAF operations and likely to come under threat.

Problem

While the importance of communications capabilities is clearly recognized, the MAF is not in a strong position to provide multichannel communications equipment and the expertise to operate it because other parts of the USAF are responsible for such capabilities. In addition, the USAF chose to divest communications expertise and instead rely heavily on contractors.[55] AMC recognizes that the MAF will need improved communications kits, particularly for command centers on the ground, but this will remain new territory for the MAF in the near term.

Similarly, AMC does not have the lead for ensuring cyber defenses. Thus, another question regarding the design of the C2 FE is how the expertise and capabilities for cyber defense will be included in the C2 FE, as it is currently not organic to the MAF.

Implications

For many years, AMC has recognized a shortfall in BLOS communications capabilities, which motivated investments in systems such as real-time information in the cockpit.[56] AMC will need to invest in the communications kits to enable ACE operations. In addition, it will need to ensure that the ACE C2 FE has the required expertise to operate that equipment. Similarly, AMC will need to ensure that the cyber defense expertise and capabilities are well integrated into the C2 FE.

The problem is larger than AMC, in that all DoD communications systems face some form of disruptions threat. Some of the response has focused on returning to high-frequency radio systems, which had been phased out in recent years.

Command of Disaggregated Forces

Activity

Another unique challenge facing a lead wing is the need for C2 over disaggregated forces. All forces under the expeditionary wing command will need to have a clear understanding of the commander's intent, as well as information about the conditions under which delegated

---

[55] Program Budget Decision-720 implemented a series of budget cuts between 2006 and 2011, among which were substantial cuts to USAF communications specialists (Thompson, *Organizing for the Future*).

[56] Oppelaar, "C4ISR Concerns for Air Mobility Command."

authorities or CBAs apply and expectations about how they should be implemented. An AEW commander may be delegated the authority to move forces within a group of airfields. This could include moving aircraft among airfields, moving support assets, or deciding on the basis of previous attacks or perceived future vulnerability whether or not to abandon a location.

Problem

A consequence of disaggregated forces is the reduction or elimination of the ability to have face-to-face contact with subordinates on a regular basis. This could increase the likelihood of misunderstandings and erode some of the interactions that lead to increased trust, which is an important aspect of command relationships.[57] It could also have other practical implications. Inherently it could introduce information latencies that make it hard for the wing command to know the precise readiness of the force at any point in time. It might make it harder to convey a commander's intent. Disaggregation of the force will make it more resilient to attack, but also could make command relationships more complicated. For instance, wing commanders might sensibly delegate control of disaggregated forces to a nearby wing in the event they lose communication with the wing command. But those conditions might change frequently, causing uncertainty for all concerned.

Implications

Commanders who are to lead MAF C2 FEs should be trained to convey their intent in clear language and to define and convey delegated authorities in ways that airmen understand. One important aspect of this training is to help leaders understand how others interpret their orders and clarifications. This is particularly important in the context of disaggregated forces, which can lead to misunderstanding due to reduced interactions.

In addition, commanders should be trained to assess and direct the disposition of assigned wing assets among multiple airfields.

Organizational Considerations of C2 FE Design

The increased demands on the C2 FE stem from the complexity of ACE operations and the disruptions caused by adversary attacks. We will consider three aspects of organizational design, which draw from current organizational theory.[58] These are the division of labor, the level of hierarchy, and coordination mechanisms.

---

[57] See, for example, Wilson, Straus, and McEvily, "All in Due Time."

[58] Snyder et. al., *Assessment of the Air Force Materiel Command Reorganization.*

Division of Labor and Expertise

The USAF already has a long-standing functional organizational structure. Consequently, in the face of new demands, the first instinct is to see them within the framework of that structure. As described earlier in the report, USAF wings are currently organized into four groups, with each of the groups containing squadrons (even with the merged MG FEs, the squadron structure remains intact). The additional demands from ACE operations align well with these squadrons, albeit with a few exceptions. Table 2.3 show how the demands are matched with the current relevant organization.

Table 2.3. Aligning Tasks and Organizations

Increased C2 Demands	Relevant Squadron
Operational planning	Not aligned
Mission planning	Operations
Recovery	Civil engineer (partial)
Logistics	Logistics readiness
CC&D	Not aligned
Intelligence	Intelligence (can be either operations group or operations-support squadron)
Communications	Communications
Disaggregated forces	Wing/Squadron commander and command staff

There are two areas that do not align with current USAF wing structure: operational planning and CC&D. Operational planning is ultimately about formulating and directing a mission. It is not currently done by operations groups or squadrons, who could be trained to perform that function, but it is distinct from mission planning, which includes a greater level of detail about each mission assigned by operational planning.

CC&D will be much harder for a wing to implement given the current expertise and organizational structure. AMC could choose to create a new force structure to organize and develop these skills. Alternatively, AMC could work with those responsible for CC&D within the USAF to develop training and tactics, techniques, and procedures (TTPs) that could be implemented by a deployed wing.

Recovery is an area that only partially aligns with the current structure. Although the civil engineer squadron has relevant capabilities in its engineer and firefighting elements, a major attack will require a more extensive coordination of activities that is not present in current wing design.

Level of Hierarchy

A wing is a functionally aligned organization.[59] Each squadron is responsible for a discrete set of tasks and for collaborating with one another. The function of the wing leadership within the organization is to ensure strong collaboration and cooperation across functional units. It also serves to mediate between the higher headquarters and the wing. This involves two-way information-gathering and transmission. The wing leadership receives taskings and the commander's intent from higher headquarters. It is then responsible for transmitting the commander's intent to subordinates and ensuring the readiness that will enable the wing to meet its tasking. Wing leadership is also responsible for gathering information from its forces to report on the readiness of the wing and the installation, as well as on completed missions, to higher headquarters and to request support.

In developing FEs, AMC will be deciding what support the wing commander needs in order to implement these enhanced wing leadership tasks, and what functions will remain the responsibility of the MG and air base squadrons that are attached to the C2 FE.

Coordination Mechanism

In considering what needs to be in the C2 FE, the functions could be assessed in terms of whether they need to decide something affecting more than one group, coordinate within a group, coordinate outside the group/wing, or report. If a function is to coordinate within a group or to report, it can remain at the squadron, or group level. Those functions that require coordination outside the group or wing and those that are geared toward decisions that affect more than one group need to be at the wing level. Operational planning and recovery are two areas that meet this criterion. Mission planning can remain at the squadron level to be effective. Logistics can be done at the squadron level in the logistics-readiness squadron; however, the likelihood of supply or lift shortages in a complex future fight suggests that units may be required to prioritize their supply requests, which is something that would need wing-level expertise.

Additional Implications for Mobility Air Force Wing C2 FE

Wing-level, adaptive cluster, and regional organizing options for deployed MAF forces will be discussed in the next chapter. Regardless of the option, MAF wing-level and below C2 FEs must be trained and equipped to connect and interact with the two distinct chains of command, that of the 618th AOC and that of the regional command (described above) for both tasking and control and reporting of mission execution. Further, if AOC responsibilities—whether those of the 618th AOC or of the regional AOC/AMD—are delegated to the MAF wing level as an

[59] A functional structure aligns related knowledge and skills together in the same unit. It is in contrast to a divisional structure, which is organized along multiple product lines. Each division contains the expertise to produce its own product line. Snyder et. al., *Assessment of the Air Force Materiel Command Reorganization*, p. 26.

adaptation to loss of connectivity or other disruptive adversary actions, the wing may require skip-echelon authority to use the next higher headquarters as an alternative and/or to establish direct user interface(s) in order to perform mission planning, prioritization, intelligence support, and flight-following functions that are normally centralized. These considerations have implications for the size, capabilities, composition, and training of MAF C2 FEs for ACE operations in an adaptive base cluster, whether located at a hub or spoke.

Chapter 3. Force Element Options

The MAF already has a capability to design and field a tailored AEW command element. The plan to adopt an Air Force force generation, or AFFORGEN, model provides a major difference in that it could allow a C2 FE to receive training specifically designed for its task and then serve as a ready force during the last phase of that cycle. In this model, the MAF needs to define the scope of the command element in advance, because that will govern the training and composition of the C2 FE. Part of scoping a C2 FE centers on whether the C2 FE should be designed for a single function, as are the majority of MAF in garrison wings, or whether it should combine functions. In the past, forward AEWs have typically involved multiple types of aircraft. Another question is whether the MAF should design wings to operate independently or whether falling in on CAF forces will serve the range of needs.

In this chapter, we develop and compare alternative organizational scoping options for MAF C2 FEs and identify the advantages and disadvantages of each. We also build on the findings from the last chapter, which distilled the additional burdens that ACE and adversary operations could place on future command elements, and consider whether these new burdens align with current wing organization. We then assess the different scoping options.

Scope of Responsibilities

RAND developed different organizational scopes by varying the MAF core functions addressed and the assumptions about whether the organization would operate on its own or combine with a CAF unit. In developing the options, we sought to achieve clear distinctions, while recognizing that numerous subvariants are also possible. The three options are an All-MAF-Capable C2 FE, a Single Function–Capable C2 FE, or a CAF Embed–Capable C2 FE. Figure 3.1 depicts these three scoping options, which are designed for a hub or major base. The figure indicates that and illustrates that these units will have echelons above them (global, theater, and possibly regional in the form of an AETF) and could also have echelons below them operating from a spoke or a cooperative security location (CSL).[60]

[60] AFDN 1-21, *Agile Combat Employment*, defines a CSL as an enduring location that has minimal U.S. presence and is maintained by contractors or host-nation support.

Figure 3.1. Hub C2 Scope Options

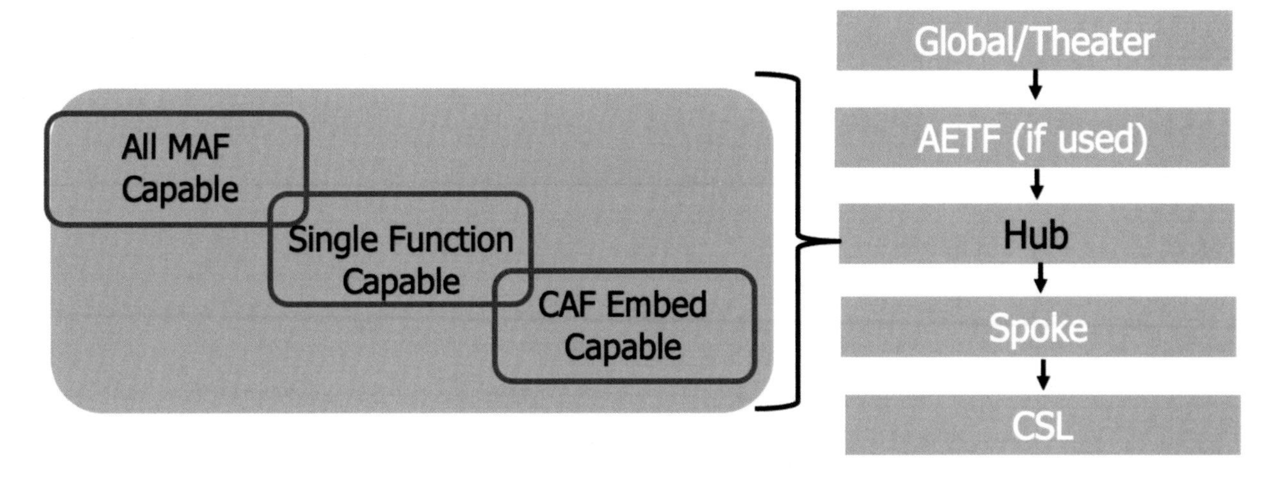

NOTE: Two of these would represent MAF-independent units: All-MAF-Capable and Single MAF Function–Capable. The third, CAF Embed–Capable, is designed to attach to a CAF unit.

All–Mobility Air Force–Capable C2 FE

An All-MAF-Capable C2 FE would have the training and expertise for effective C2 of all types of MAF missions, including both air-refueling and airlift operations. Multiple MG FEs (that is, aircraft and maintenance detachments) could be attached to the lead wing C2 FE, with each one bringing weapon system–specific expertise to augment the C2 FE. They offer the possibility of commanding MAF clusters in ACE operations, which could be particularly important if PACAF and USAFE intend clusters to be under the command of a single wing. The C2 FE would be expected to have personnel with a mix of expertise and with cross-training to support the commander in leading a multiplatform MAF installation or cluster. Since MAF units are typically organized in homogenous wings, this would require adjustment to personnel and/or training.

Mobility Air Force Single Function–Capable C2 FE

Currently, MAF forces are mostly organized in single-mission garrison wings, which are designed to train the force, but not specifically for wing leadership to deploy as a command unit.[61] By designating C2 FEs for some of these units, the MAF could have ready mission expertise postured to C2-deployed units for each type of MAF force. They also offer the possibility of commanding homogenous MAF clusters in ACE operations. While the MAF more typically operates in AEWs that include CAF forces as well, an MAF-only cluster could be desirable under certain conditions (to take advantage of locations not suited to or available for the CAF).

[61] The 60th AMW (Travis) and 305th AMW (McGuire) have both airlift and tankers, and transitioning tanker wings will be dual KC-135/KC-46 wings for a period. While not functionally different, they are different MDS.

Furthermore, there is a precedent, in that MAF forces operate without the CAF in support of humanitarian operations.

Combat Air Force Embed–Capable

In recent conflicts in Iraq and Afghanistan, AEWs have tended to be co-located units, or wings with both CAF and MAF forces. At locations such as Bagram Air Base, Afghanistan, and Joint Base Balad, Iraq, large AEWs comprised of both CAF and MAF assets conducted air operations under C2 of the theater command, with all forces under the tactical control of the AOC. In the future, this command structure could be retained. Alternatively, MAF units embedded within a CAF-led composite wing could be controlled by the CAF wing itself in a direct-support role. These AEWs would need MAF expertise in the wing leadership in order to effectively employ MAF assets. In the same way that an AMD provides MAF expertise in the current regional AOC, a CAF Embed–Capable C2 FE could provide relevant expertise to future lead wings.

How FE Scope Aligns with Mobility Air Force Priorities

In considering the scope of future FEs, the MAF will consider balancing a number of goals and constraints. In this section we assess how each of the MAF C2 FE options might meet different high-level MAF goals. The purpose of developing the FEs is to be better prepared for conflicts involving attacks on airfields, so we consider attributes considered important for operating in future environments. In these environments, the MAF will continue to perform its core functions, so the support it provides and the efficiency of its use are also important factors to consider. Further, since communications is such a fundamental aspect of modern command, we consider differences in communications demands. We then summarize our assessments of the options against these considerations.

Resilience to Missile Attack

USAF agility concepts seek to increase the ability of the force to disperse, maneuver, and recover the force. The different MAF C2 FE scoping options may have some inherent differences in terms of their ability to support greater dispersal. Focusing first on an illustration from a single geographic area, let's imagine two CAF wings, each operating in separate hub-and-spoke clusters. In proximity to these two clusters are a number of other joint and coalition users. With a CAF Embed–Capable C2 FE, the MAF could place one unit at each of the two CAF hubs. However, we assume that each CAF Embed is a single MDS unit, so depending on the demands, there might be a requirement for more than one embedded FE within each cluster. With additional aircraft placed at the CAF locations, each one becomes a more lucrative target; meanwhile, the MAF has not added any new locations to increase the dispersal of the force.

Alternatively, in the situation outlined above, if the MAF were given access to another location, it could place the All-MAF-Capable C2 FE at that location and size their footprint to support the two CAF clusters, as well as the joint and coalition users. In that case, the USAF is more dispersed in that geographic area because the MAF has added another location. If, instead, the MAF is given two additional locations to operate from, it could operate a Single Function C2 FE unit, one at each location. Still, this added resilience comes with a cost of the additional personnel and equipment needed to operate these locations.

If the MAF can operate from locations that are not suited to the CAF or that partners do not make available to the CAF, that could improve the overall resilience of the USAF. One type of airfield that should be considered is that of the commercial airport. In the past, host nations have not given uniform access to the USAF to operate, but instead place various conditions on access for combat aircraft, in particular.[62] Permitting MAF forces to operate from commercial airfields and/or use their resources could increase the overall dispersal of USAF forces. On average, airports are larger than military air bases, and so could require more missiles to attack, which further adds to posture resilience. Certainly, the larger two C2 FEs could operate independently from an airport. The CAF Embed–Capable C2 FE is not designed to operate independently.

Support to User

We assume that combined planning is an important part of MAF effectiveness and the ability to support users. The scope of the C2 FE is relevant to the effectiveness of the support in situations in which communications disruptions or attacks on C2 nodes force the unit to operate separately from the AOC. Combined planning is particularly important for the design of air-refueling support, which is extremely sensitive to operational decisions about distances, aircraft density, and the expected threat environment. Sequential tanker planning, after CAF missions have been planned, may require repeated iterations or difficult late adjustments. The CAF increasingly conducts planning in sensitive compartmented information facilities (SCIFs), which may exclude some MAF participation due either to space and equipment limitations or to information restrictions of the SCIF.

As the USAF employs more agility concepts, the need to link CAF planning and airlift planning will grow. When CAF missions utilize several airfields to complete a mission, instead of the single airfield missions of the past, those additional locations will typically need airlift support. Part of our assessment of effectiveness is the assumption that the proximity of planners will help ensure needed support. Problems could arise in supporting such agility missions when the additional airfields extend beyond the range of MAF airlift units assigned to support the originating CAF locations. Or, even if airlift aircraft could travel those distances with the CAF aircraft/units they are supporting, the opportunity cost might be too great. (Having the tankers utilize their cargo capacity to support such operations could be another option.) In supporting

---

[62] For instance, partners historically have not placed restrictions on airlift operations, while there are instances of restrictions on combat operations and strike operations in particular (Pettyjohn and Kavanagh, *Access Granted*, p. 76).

that single mission, many other users could face interruptions in their airlift support. In those situations, lateral communication with airlift units assigned to support the distant locations could ensure the support, or the local airlift commander would have to make the decision to balance the needs of all assigned users.

The MAF supports joint and coalition users. Planning for common user support comes through the JDDOC to the AMD. In situations where an MAF unit will provide support to a joint user for an extended period—for instance, when supporting an air maneuver of an Army unit—the support will benefit from an ability to provide liaisons to that unit. Other, more routine support does not rely on such direct planning. For sustained operations in Iraq and Afghanistan, the supplies for common users could be predicted well in advance, and therefore the airlift support could be scheduled in advance. Similarly, the aircraft missions could be scheduled in advance, and tanker support could be planned accordingly. There was little need to react quickly to schedule and launch new missions in response to adversary surprise. In a future conflict with Russia or China, the demands will be less predictable and more dynamic. Currently, any unexpected changes in user demand are expected to be managed by the AOC/AMD. However, this will be no easy task at the theater level. Thus, in the future, ways to overcome information saturation at these centralized nodes should be explored.

All three of the C2 FE scoping options can be effective in supporting users, but each one provides utility in a different way. The CAF Embed–Capable FE option can be very effective in supporting one user. That effectiveness is predicated on an ability to be well integrated into the planning of the host unit. For tanker units, that would involve being embedded in the AEW's mission-planning element. In cases where the MAF forces can embed with the user planners, that will provide the best understanding of their needs and priorities. It will also ensure timely updates. If that is not the case, some of the benefits of tethering are eroded.

The All-MAF-Capable and Single Function C2 FEs are better suited than the CAF Embed–Capable C2 FE to support multiple users, particularly if employed as stand-alone units. This potentially exacerbates the planning challenges discussed above, as the more users a unit supports, the greater the barriers to cooperative planning. However, independent units will likely have more aircraft, which may allow them to better adjust MAF support among users as their needs change over time. To the extent that MAF units can serve a larger number of users, the need for slack capacity across users is reduced. The closer that MAF assets are tied to specific users, the better they will understand and be responsive to those needs, but this scenario also assumes a future in which the MAF C2 FE has a role in prioritizing missions. This is certainly not a settled matter within the USAF, but in light of the complex operations and the tremendous friction anticipated in future conflicts, new operating concepts are being developed to give more authority to lower echelons to cope with unexpected changes. Even in situations in which the AOC/AMD is fully functioning and communications channels are working, instead of tasking MAF assets to full capacity, a system designed to allow some slack capacity and lower-echelon initiative might better meet emergent support requirements.

Utilization of Mobility Air Force Capabilities

Utilization of MAF assets is the other side of the coin. While the previous section assessed how the different scoping options are suited to supporting the user, this one focuses on optimizing the utilization of MAF aircraft. Also, it is motivated by anticipated future scarcity, such that joint demand for airlift and air refueling exceeds capacity.

A CAF Embed–Capable C2 FE option may require more MAF forces overall. The daily MAF support requirements are not constant, as the pace of operations changes, and the needs of users ebb and flow. By assigning MAF assets to a single user, the MAF would need to provide enough capability to meet peak demands, which means that at other times the MAF assets will have slack capacity. (Alternatively, they could be sized to meet steady-state demands and seek augmentation to meet peaks.) Such a unit is likely to be placed under the command of the supported unit. Although this type of consolidated unit has been typical in recent conflicts, those installations were typically large and included senior MAF representation. A CAF wing with a small MAF embedded unit may face a learning curve in terms of the tasks that fully utilize MAF support.[63] This certainly could be addressed through training in the future.

Communications

The CAF Embed–Capable C2 FE would not be subject to communications disruptions with the users(s) they are embedded with. All the options have a similar reliance on long-range communications to AOCs and other forces. If LOS communications are feasible (they are hard to disrupt from a distance), then all the options could communicate, at least partially, with users within LOS range, provided the units have the needed equipment and expertise.[64] While in the past the MAF often operated with CAF forces that provided the communications capabilities, new investments in communications equipment and personnel may be needed for each FE to be effective.

Summary of Advantages and Disadvantages of Different C2 FE Scopes

In the following section, we discuss the advantages and disadvantages of the three options designed for hub-sized employment. Table 3.1 summarizes the assessment of the three C2 FE designed for hub-sized responsibilities. The end of the chapter identifies and assesses options for smaller-force packages designed to operate from spokes within a cluster.

---

[63] For instance, in a recent exercise, MAF forces placed under the command of a CAF wing were ordered to sit on alert and thus missed an opportunity to provide an airdrop to a simulated ground user (39th Airlift Squadron, "After Action Report for MOSAIC TIGER").

[64] Kallberg and Hamilton, "Resiliency by Retrograded Communications."

Table 3.1. Qualitative Assessment of Options

Quality	All-MAF-Capable	Single Function–Capable	CAF Embed–Capable
Level of dispersal (resilience to missile attack)	Consolidation results in fewer overall hubs, but stand-alone MAF increases resilience	Potentially allows most MAF hubs; increases resilience	If co-located, MAF forces increase value of attacks on their locations
Support to user	Independent MAF unit can be flexible in supporting multiple users	Independent MAF unit can be flexible in supporting multiple users	Best understanding of co-located users; potentially less able to support joint or coalition users
Utilization of MAF aircraft	Best positioned to prioritize requests across multiple users; efficient use of assets to support all users	Positioned to prioritize requests across multiple users; efficient use of assets to support all users	Dedicated to one user; less efficiency if CAF not able to fully utilize
Comms reliance/ fragility	Without communication with users, support feasible but less efficient	Without communication with users, support feasible but less efficient	Not reliant on comms with co-located user

Advantages of All–Mobility Air Force–Capable C2 FE

An All-MAF-Capable C2 FE provides the MAF with the most flexibility when employing the force. Major contingencies are likely to feature all types of MAF forces operating in close geographical proximity, and this C2 FE provides a ready option for efficient C2. This C2 FE could be sent to a single installation, or cluster of airfields, and be prepared to C2 a variety of MAF MG and support elements. There would be no need to reserve installations or clusters for just one type of MAF force. The forces could be tailored to balance the infrastructure available, the threats they face, and the users they support. There is some historical justification for the MAF to be concerned about the basing access it will be provided in a future conflict. For instance, in the Gulf War, COMALF repeatedly argued for more in-theater basing for airlift. A postwar survey done by the predecessor to AMC highlighted this problem.[65] Co-locating multiple MAF force-generation elements under the command of an All-MAF-Capable C2 FE potentially reduces demand for MAF basing space.

A stand-alone MAF wing provides a better ability to balance and serve the needs of multiple users than an embedded element. The MAF has multiple users, including the CAF, joint users, and coalition partners. The demands from these partners will change frequently. An independent MAF wing is in a better position to balance those needs and more efficiently support multiple users.

Users might find it easier to interface with a single MAF-wide-capable C2 unit, rather than individual C2 FEs focused on a single MAF mission. Similarly, AMC would have maximum

[65] Devereaux, "Theater Airlift Management and Control—Should We Turn Back the Clock to Be Ready for Tomorrow?" p. 33.

flexibility to adjust the right mix of airlift, aerial refueling, and aerial port C2 elements to meet the uncertain future needs.

The MAF would have the option to tailor the C2 element by stripping away capability in situations in which the AEW does not need the full set of MAF functions. In terms of readiness, shedding assets would be preferred to adding new expertise on an ad hoc basis, which would be the case with the other, more tailored options considered below.

Disadvantages of All–Mobility Air Force–Capable C2 FE

MAF missions and tasking are distinct and may not be mutually reinforcing. Although aerial refueling and airlift are grouped together by the USAF under AMC, there are certain aspects of their mission sets that are quite distinct. Like the fighters and bombers they support, the aerial refueling missions tend to get planned in the AOC's combat plans division. In contrast, while airlift units might be supporting some of the same forces as the tankers, they transport joint user cargo and personnel to and between deployed operating bases, and their mission taskings originate at USTRANSCOM or the JFC's JDDOC. The aerial port function is also quite distinct, as it focuses on the receipt and transshipment of cargo. As a result, there may not be a lot of benefit to combining airlift and refueling MAF functions under one C2 FE. Often an organization dedicated to one thing can achieve better results than another organization that tries to do multiple things.

The All-MAF-Capable option will likely have the largest footprint. This could be a disadvantage in that the potential impact of attacks on the airfield makes a streamlined force desirable to reduce exposure to losses. However, this might not require the most personnel overall across a geographic area, when AMC considers the total force devoted to creating these new C2 FEs. For instance, AMC might determine that it can field fewer All-MAF-Capable C2 FEs, compared with the total number of single-mission, or other tailored options.

Advantages of Mobility Air Force Single Function–Capable C2 FE

The MAF is already organized along functional lines, so a single-function C2 FE provides the least disruption and the clearest organizational continuity with the peacetime structure. It also means that the leadership of MAF wings will consistently offer a high degree of experience and expertise in their mission area, which may not be achieved if wings combine multiple missions.

This option also provides a stand-alone MAF wing, which better allows the MAF to serve the theater/global needs of users for the specific function, compared with the more limited capability below. The MAF has multiple users, to include the CAF, joint users, and coalition partners. The demands from these partners will change frequently. An independent MAF wing is in a better position to balance those needs than an embedded option.

Disadvantages of Mobility Air Force Single Function–Capable C2 FE

A single-function capability may create situations in which two MAF-wing C2 FEs are either co-located, or assigned to nearby airfields in order to have the right C2 elements and forces in a given area. AMC may need to create more of these than would be the case with the All-MAF option above. In the Pacific, USAF forces will be aggregated in a few desirable areas. Putting two or more MAF C2 FEs in confined areas may be difficult, in that each location represents somewhat of an opportunity cost in a scarce environment. In the past, the USAF has experienced challenges when two separate commands are co-located. For instance, a USAF study of airlift in Vietnam found that "duplication and/or overlap of the responsibilities and functions occurred in aerial ports, [and] airlift control elements. . . . In this case there were two airlift forces with similar capabilities performing within and between an area command."[66] In the Gulf War, the King Khalid Air Base in Saudi Arabia had both an airlift control element and airlift wing operating independently, which resulted in "constant territorial disputes involving command and control authority, ramp space, MAPS (Mobile Aerial Port Squadron) support, and decision-making authority."[67] If the MAF chooses to implement Single Function–Capable C2 FEs, in situations where two or more C2 FEs are co-located, they might consider a policy of naming a single overall commander.[68] However, it may not always be the case that the MAF forces are under the same commander, such as if an airlift unit operating under the JFC's operational control (OPCON) is co-located with an AMC aerial port unit under the OPCON of USTRANSCOM. Those forces might benefit from leadership guidance that clearly delineates their respective roles and resources.

Advantages of a Combat Air Force Embed–Capable C2 FE

A CAF Embed–Capable C2 FE has several advantages. It offers a tight coupling between the CAF and MAF forces in each cluster. Co-locating command elements can be expected to provide the best alignment between the CAF missions and MAF support. They will benefit from unity of command at the wing level.

This option would have the smallest individual C2 FE footprint for the MAF, as it would provide only MAF-specific force employment and mission-planning expertise. All other C2 functions would be handled by the CAF or the installation authority.

Disadvantages of Combat Air Force Embed–Capable C2 FE

Joint and other users may get reduced support depending on how much of the total MAF force is embedded within composite wings in a direct support role. Alternatively, disaggregation to support each CAF unit in this way may require more MAF force structure to maintain its current level of support to the joint force while simultaneously providing dedicated support to the CAF. Table 3.2 summarizes the advantages and disadvantages of all three options.

---

[66] Carter, "Theater Air Mobility," p. 29, citing Betty R. Kennedy, *Air Mobility En Route Structure: The Historical Perspective, 1941–1991*, AMC Office of History, Sept 1993.

[67] Devereaux, "Theater Airlift Management and Control," p. 34.

[68] As Hap Arnold colorfully quipped, when two organizations are joined, "unity of command can be expressed only by a superior" (quoted in Allard, *Command, Control, and the Common Defense*, p. 102).

Table 3.2. Hub C2 FE Scoping Options, Advantages and Disadvantages

MAF C2 Options	Description	Advantages	Disadvantages
All-MAF-capable	C2 of all MAF forces MAF AEW	Most flexibility for AMC Option to tailor by removal Best ability to serve all users One stop for users to interface with MAF May allow fewest total C2 FEs	Largest footprint Multifunctional organization more cumbersome than single-mission org
Single function–capable	C2 of MAF AEW cluster	Least change transitioning from garrison to lead wing Ability to serve theater/ global needs	May require multiple MAF wings to operate in the same cluster or subregion May require more total C2 FEs
CAF embed–capable	MAF planning for wing cluster	Tight coupling of CAF/MAF Smallest C2 FE	Disaggregation to support CAF may increase total MAF C2 FEs required MAF has others to support

Mobility Air Force Command at Spoke Locations

One major USAF adaptation to the threat of missile attacks against airfields is to disperse the force. In forward areas, this could result in small detachments of aircraft operating from a variety of locations and conditions. As discussed above, wing commanders should be prepared for C2 forces to be operating from multiple locations simultaneously.[69] However, just as a wing might have to operate on its own, so might a spoke if it becomes disconnected. What is the right C2 footprint for spokes? The answer to this question is interrelated with the design of the wing C2 FE. A small spoke command element puts more of a burden on the wing, whereas a large spoke's C2 FE would be less reliant on reachback. In this section, we consider the command needs of such locations. The MAF could consider scoping a spoke C2 FE in two different ways. Figure 3.2 illustrates the scoping options.

Mobility Air Force Spoke C2 FE

The larger MAF Spoke C2 FE would include scaled-down wing command elements at each location. This would allow the spoke an ability to operate autonomously if its parent wing gets attacked. If there are MAF aircraft based at the spoke, there should be some sort of a C2 FE who is dual-hatted as the spoke commander, and who has a small staff consisting of intelligence; tacticians and mission planners; a small team of combat control, contingency response,

[69] "Forces must be able to rapidly execute operations from various locations with integrated capabilities and interoperability across the core functions" (AFDN 1-21, *Agile Combat Employment*, p. 5).

multicapable airfield operations types, and transient alert type maintainers; and a minimum of two personnel providing round-the-clock communications support. If there are no MAF aircraft based at the spoke, a C2 FE is not needed, but a command post function would be useful to relay orders and communications.

Figure 3.2. Spoke C2 Scope Options

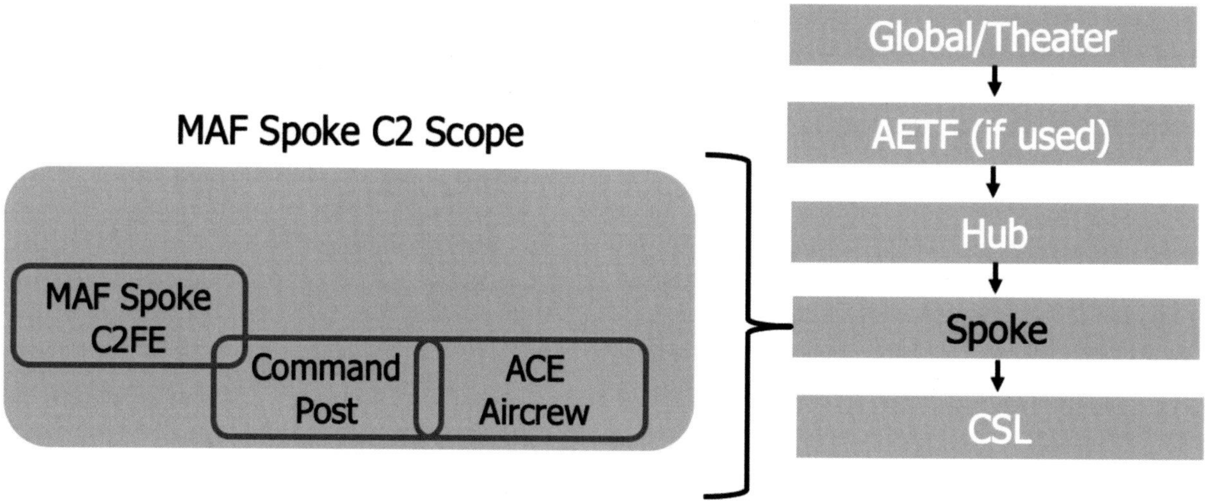

MAF Spoke C2 Scope

MAF Spoke C2FE

Command Post

ACE Aircrew

Global/Theater

AETF (if used)

Hub

Spoke

CSL

NOTE: CSL = cooperative security location.

Mobility Air Force Command Post C2 FE

A second option is for each spoke to have a very lean Command Post C2 FE, with communications equipment and expertise and a small staff component. In this configuration, the spoke would be relying primarily on reachback to a wing C2 FE. In situations where they become disconnected, the detachment commander would be expected to direct the employment of the force, consistent with the commander's intent and delegated authorities.

If a spoke is very distant from its wing hub in a far-forward area, it will face a chaotic environment. If the spoke is operational, it should be prepared to receive and turn aircraft that are not assigned to the spoke. In forward areas, aircraft may face emergencies that force them to land at the nearest available airfield. The benefits are provided not only to aircraft facing emergencies; if it was a known asset, aircraft could plan to refuel as part of a complex mission from a distance. In this regard, a spoke with an ability to turn all types of aircraft could be an important asset for the joint force. An ability to communicate with higher headquarters could be another service to those drop-in aircraft, whose crews could use the communications capabilities to communicate directly with their leadership to confer about mission taskings or discuss possible maintenance problems. This is another example of the benefits of breaking the tight coupling of the concept of command with forces at specific locations. It envisions a small number of forward-operating locations that meet the needs of multiple units.

Spoke Contribution to a Modular Force

Either of the two spoke options merits consideration for its ability to facilitate dispersal of MAF forces and to effectively C2 them. Another feature that might make them attractive is their ability to form the smallest C2 FE that could allow the MAF to have more flexibility. We assume that many forward areas will need MAF support, but in many cases this will be fleeting, or will require fairly modest footprints because the footprint of the supported forces will also be small. In these situations, the FEs sized to operate from a hub would be too large. A spoke-size force might be well suited to operating in such an environment.

A spoke-size MAF C2 FE could play the role discussed in this report for embed units. One of the key differences might be the need for communications capabilities at a spoke, which would not be necessary when co-located with a CAF unit, but the size and the functions envisioned for these elements might be close enough that they could be structured to play either role.

Finally, if a spoke C2 FE were combined with a Single Function–Capable C2 FE, it could achieve the intent of the All-MAF-Capable C2 FE.

Advantages of Mobility Air Force Spoke C2 FE

Re-creating a wing C2 FE in miniature at each spoke provides the MAF with the most capability and the greatest resilience to attack. Each spoke would be able to operate independently, if necessary. This arrangement reduces the need for reachback. It also reduces the value of attacking wing hubs because the spokes can continue operating. It would allow aircrews to focus on doing their mission, instead of spending time gathering information and trying to figure out what their mission should be, as might be the case with the second spoke option described below.

Disadvantages of Mobility Air Force Spoke C2 FE

This would have a larger footprint compared with the alternative, which means the number of spokes that the MAF could support would be fewer. Of course, as the footprint of a location shrinks, the added value of a robust C2 FE may also shrink.

Advantages of a Command Post C2 FE

A Command Post C2 FE has a lean footprint that features a spoke commander and a command post to provide the necessary communications equipment to allow reachback. This small footprint means the MAF can create multiple Command Post C2 FEs to facilitate greater base dispersal. It leverages the wing-level C2 capabilities, thus allowing the spokes to focus on MG and leveraging some economies of scale.

Disadvantages of Command Post C2 FE

If communications are degraded, the spoke may not function as well as the spoke C2 FE alternative. Aircrews may have to take time to gather information to inform the mission planning, whereas a dedicated command element could do that for them, potentially saving time and achieving less disruption to sortie rates.

Chapter 4. Findings and Recommendations

The MAF already has many of the qualities the USAF is looking for in developing a more agile force. It has experience operating small detachments for long periods of time, sometimes far away from their home stations. But the MAF currently lacks C2 FEs that are trained and ready to deploy and operate as a unit in a complex future operating environment employing the USAF's new agility concepts and CBAs. Defining such FEs and organizing and equipping them could be an important step for the MAF to be able to provide highly dispersed forces prepared for disaggregated air operations, which will make them more resilient and dynamic and better able to support the joint force in order to prevail in the new operating environment.

Key Findings and Recommendations

Does the Mobility Air Force Need Independent C2 FEs?

The MAF can operate from locations that are not suited to the CAF or that partners do not make available to the CAF. Host nations have not given uniform access to the USAF to operate from commercial airfields in the past, with prohibitions on combat operations being the usual case. An independent MAF unit could utilize such locations and operate effectively within the broader C2 structure of USTRANSCOM and/or the theater. Utilizing locations that would otherwise be unavailable improves the overall resilience of the USAF. Dispersal is a key aspect of future operating concepts and is associated with efforts to increase the resilience of USAF forces to adversary attacks.

The C2 elements designed for future conflict will need to account for geographic separation of the force under threat of dynamic physical attacks and cyberattacks. Command elements at hubs will need to have the capacity to monitor the airfields and forces within their cluster, allocate cluster resources among locations, communicate with higher headquarters and directly with users, and direct the movement of aircraft and capabilities. Spoke locations may also need enhanced C2 capabilities. An ability for the MAF to operate independently (not attached to another unit) from forward airfields contributes to enhanced resilience of the force. Enhanced C2 capabilities at the unit level increase the overall resilience of the C2 system—whether the theater C2 system or the global USTRANSCOM system.

Recommendation: For future high-end fights, the MAF should develop a wing-level capability for C2 of independent MAF forces operating from multiple locations (e.g., hub and spokes).

What Scope?

In a future conflict, the MAF will be asked to conduct global operations, and MAF forces will operate from many kinds of locations, with different threat profiles and different configurations of MAF forces. Therefore, defining generic FEs that will meet this wide range of needs will be difficult and perhaps not productive.

A modular approach maximizes capacity, redundancy, and flexibility to meet a range of needs and simplifies force presentation. A modular MAF C2 FE provides the basis to employ capability at multiple echelons: spokes, hubs, and AETF. For example, depending on adaptive base cluster composition (number of hub and spokes, size/distance, and number/types of aircraft assigned), two or more modular MAF C2 FEs could be deployed in combination to C2 a hub with multiple spokes or to perform regional MAF C2 functions on an AETF staff. A single MAF C2 FE would typically be capable of performing envisioned MAF C2 functions for MAF forces deployed to a spoke location.

Recommendation: The MAF should adopt a modular approach to C2 FE design to better meet the expected range of future demands within the force structure available. A modular approach could maximize capacity, while providing flexibility. For instance, Mobility wings could build small C2 FE cells consisting of the following numbers of personnel: a mission planning cell of two tacticians and two flight planners, one for intelligence, two for current operations, two for maintenance operations, one for communications, and a C2 FE commander. Based on need, these smaller C2 FE cells could be merged with others within the Wing and eventually with those from other Wings to scale up the C2 FE to an appropriate size based on its scope of responsibility. As the C2 FE grows, additional functions—such as air traffic control, host nation relations, or security forces—can be added as needed.

What C2 Adaptations Are Implied by Adversary Threats to Centralized C2?

Airpower is most effective when centrally controlled. Airpower should be centrally controlled at the highest level feasible. While centralization in recent conflicts has been feasible at both the regional AOC and globally at the 618th AOC; the USAF should prepare for situations in which centralization at those levels is not feasible. Whether that should occur at an AETF or at a wing is something that needs careful consideration.

The C2 implications of potential adversary attacks on AOCs and communications capabilities are far-reaching. These could include degraded or disrupted communications; fragmented chains of command, due either to attacks on command nodes or to network degradations; data integrity loss, due to cyberattack; and logistics scarcity, due to attacks on logistics nodes. Communications disruptions and fragmented chains of command could force wing commands or lower echelons to operate autonomously for periods of time. The USAF is considering how forces cut off from higher headquarters should respond. While commands could publish multiday orders to minimize the consequences of such disruptions, actions on the part of the adversary, weather, or even "normal" operational friction such as equipment breakage can force those plans to change, which, in turn, means that the autonomous elements would be out of synch with the rest of the force, if they kept repeating old missions.

The concepts and authorities governing isolated units is evolving, but it needs to evolve beyond assumptions of short duration or limited disruptions. It remains a difficult question because centralized control is so fundamental to USAF operating concepts. While logically it is understood that disruptions should be expected in the future, any adaptation would be a major change. The USAF agility concepts emphasize the need to empower lower echelons to make decisions. Preparing FEs for disruptions does not discard or break centralized control; rather, it is one element of prudent preparations for C2 resilience.

Recommendation: The USAF should continue to develop doctrine and TTPs to allow forces to centralize command at the highest feasible level in the face of attacks on C2 nodes and communications.

How Will C2 Differ in Future Operations?

The increased demands on the C2 FE stem from the complexity of ACE operations and the disruptions caused by adversary attacks. As described in earlier chapters, these could pose new command burdens related to planning, logistics, and CC&D. The call for enhanced planning deserves attention in the context of the previous recommendation regarding the centralization of command.

The problem for an MAF unit cut off from higher headquarters is somewhat different than it is for a CAF unit, because MAF forces support the CAF and other joint users. As MAF forces operate, they interact with their users. Thus, they can develop an understanding of user needs and have a means for periodic information updates of their needs that might prove to be an advantage in effectively adapting to the loss of connectivity with higher headquarters. If the MAF C2 FE needs to be prepared to conduct operational planning, that will influence the design of the C2 FE. This would drive demands on the planning staff to gather more information and different information. It could also change the way they prepare options and assess risk.

- Wing planners will need information to develop effective operational courses of action (COAs). This could increase the time spent communicating with users to identify needs and gather information, if some communications are available.
- Alternative means of gathering information will be needed in situations when communications are very degraded. For instance, gathering information from arriving aircraft might be one means of obtaining needed information. This adds time but also may introduce a need to have efficient, alternative means of gathering information from arriving aircraft.
- Planners also need to be able to formulate COAs and assess risks to aid a commander's decisionmaking. There will be less certainty about conditions and the needs of users.

Recommendation: MAF C2 FEs should be prepared to assume limited operational planning functions of higher headquarters (AOC/AMD) temporarily under CBAs. The ability for forces to implement delegated CBA is something that needs further development within the USAF.

Recommendation: Communications will be contested and are a fundamental part of C2, so the MAF needs to develop deployable communications capabilities for command elements that are suited to future operating environments and train with them regularly.

What Are the Unique Demands of the En-Route System?

The potential for attacks on AMC GAMSS nodes forces new consideration of the GAMSS C2 footprint and the authorities required for those nodes to be able to redirect flights transiting GAMSS locations as needed to respond to changing conditions and adversary attacks. The ability for GAMSS units to coordinate theater airlift and joint logistics forces in order to link GAMSS-provided cargo with theater logistics capabilities also needs further development.

Recommendation: AMC should work with the Joint Staff and regional commands to prepare for attacks on APODs and develop appropriate authorities and TTPs to allow the GAMSS nodes the flexibility to function in resilient ways and link with joint logistics capabilities.

Next Steps

In this report, we have focused on the demands on AMC forces in future conflicts and have provided a qualitative assessment of different options. An important next step will be for AMC to use this information to assess the personnel and resource implications of these different options so that an informed direction can be set. In addition, MAF C2 FEs should get training to prepare them for future operating environments and advanced adversary threats. Commanders and their staff need to prepare for new functions and stresses. We have identified areas such as enhanced planning, logistics, CC&D, and recovery operations. These represent wholly new or substantially altered functions that should be a focus for AMC training for these roles.

AMC will need to develop exercise and training to certify MAF C2 FEs to meet the demands of future conflicts, and operational readiness exercises and inspections should reflect these new demands. Based on current USAF concepts and doctrine, requirements include

- understanding of MAF-specific global/regional (or operational/tactical) COMREL for ACE by theater of employment (e.g., U.S. Indo-Pacific Command [INDOPACOM] versus U.S. European Command [EUCOM])
- familiarity with potential ACE operating locations/basing and playbooks for potential schemes of maneuver (or mission flows) by ACE theater of employment (INDOPACOM versus EUCOM), including potential AETFs, adaptive base clusters, hubs/spokes, and "beyond spoke" contingency locations
- the capability to temporarily assume limited AOC/AMD operational planning functions and user coordination under CBAs if connectivity is lost with operational C2 headquarters

- tactical mission planning expertise that incorporates increased branches, sequels, and MTO guidance
- the ability to C2 disaggregated forces (at hub/spokes in an adaptive base cluster or within a regional AETF) that are subject to missile, unpiloted aircraft, or cyberattacks
- the ability to secure resources in a communications-denied environment (e.g., push-logistics)
- the ability to provide intelligence I&W of attacks and threats to installations (hub or spoke)
- identification of vulnerabilities and implementation of CC&D mitigations
- the ability to manage complex responses to recover/operate air bases after attack
- development of modular communications equipment and expertise.

The ability for the MAF to command an ACE-adaptive base cluster independently includes not just the C2 FE, but also the operate-the-air base (OAB) functions and associated dependent OAB C2 FE to provide full lead wing functionality. MAF wings are not standardized. Thus, some types of AMC wings are suited to providing the full range of capabilities, while others will lack the OAB capabilities.[70]

Given the dynamic nature of the envisioned ACE operations, TTPs for both MAF C2 FEs and OAB C2 FEs should be developed in coordination with PACAF, USAFE, and ACC. During contingency air operations, MAF units and MAF aircrews will likely beddown at, deploy from, or transit a wide range of, adaptive basing options led by commanders and C2 teams from multiple MAJCOMs. Therefore, it is critical that operating procedures be standardized USAF-wide in order to maximize both force survivability and mission effectiveness.

Recommendation: AMC should advocate that the Headquarters Air Force designate the lead documents and concepts to which MAJCOMs should align all agility concepts and training. In absence of this guidance, individual MAJCOMs run the risk of developing their own agile concepts, which may or may not work well with others.

In forward areas, aircraft may face emergencies that force them to land at the closest available airfield. An ability for a small detachment to turn all types of aircraft could be an important asset for the joint force, coupled with an ability to communicate with higher headquarters. This capability would be a limited one due to the broad training required for a small team to support a wide variety of aircraft with limited maintenance and communications capabilities. This capability to support drop-in aircraft, whose crews could use the communications capabilities to communicate directly with their leadership about mission taskings or possible maintenance problems, could be an important joint asset in forward areas. As a joint resource, the capability should meet a common vision.

---

[70] For example, AMC has active-duty flying wings with installation command responsibilities (6th AMW, 19th AW, 22nd ARW, 60th AMW, 92nd ARW, 436th AW). These wings should be capable of sourcing both modular MAF C2 FEs and associated OAB C2 FEs. AMC "mission wings" (62nd AW, 305th AMW, 317th AW, 437th AW) contain no installation support personnel and could source only mission-specific modular MAF C2 FEs. Finally, AMC's air base wings (87th Air Base Wing, 628th ABW) would source only OAB C2 FEs.

Recommendation: AMC should work with ACC and Air Force Global Strike Command to develop the capability for agile forward bases to be capable of servicing both CAF and MAF aircraft. The Expeditionary Center could start this process by recommending a modular set of unit type codes for force providers to codify as a deployable drop-in support capability for spokes and forward operating locations.

Abbreviations

ABS	air base squadron
ACC	Air Combat Command
ACE	agile combat employment
AE	aeromedical evacuation
AETF	air expeditionary task force
AEW	air expeditionary wing
AFB	Air Force base
AFDN	Air Force Doctrine Note
AFDP	Air Force Doctrine Publication
AFFOR	Air Force Forces
AFI	Air Force Instruction
AFMC	Air Force Materiel Command
AFPAM	Air Force Pamphlet
AMC	Air Mobility Command
AMD	Air Mobility Division
AOC	Air Operations Center
APOD	aerial port of debarkation
ARW	Aerial Refueling Wing
ATO	air-tasking order
AW	Airlift Wing
BLOS	beyond line of sight
C2	command and control
CAF	combat air force
CBA	conditions-based authority
CC&D	camouflage, concealment, and deception
COA	course of action

COD	combat operations division
COMALF	Commander Airlift Forces
COMREL	command relationship
COP	common operating picture
CSL	cooperative security location
DAF	Department of the Air Force
DoD	Department of Defense
EUCOM	U.S. European Command
FE	force element
FORGEN	force generation
GAMSS	Global Air Mobility Support System
GOC	Global Operations Center
I&W	indications and warning
INDOPACOM	U.S. Indo-Pacific Command
JDDOC	Joint Deployment and Distribution Operations Center
JFACC	joint force air component commander
JFC	joint force commander
JP	Joint Publication
LOS	line of sight
MAF	Mobility Air Force
MAJCOM	major command
MDS	mission design series
MG	mission generation
MTO	mission-type order
OAB	operate the air base
OPCON	operational control
PACAF	Pacific Air Forces
PAF	Project AIR FORCE

SCIF	sensitive compartmented information facility
STAR	standard theater airlift route
TTP	tactics, techniques, and procedures
USAF	United States Air Force
USAFE	U.S. Air Forces in Europe
USCENTCOM	U.S. Central Command
USTRANSCOM	U.S. Transportation Command

Bibliography

5th Combat Communications Group, *5 CCG Planners Guide*, May 30, 2018.

19th Airlift Wing, "MAF C2 Force Element Table Top Exercise (TTX)," December 2021.

19th Airlift Wing, "ROCK-I 21-02 Lead Wing Experiment: After Action Report," June 17, 2021.

39th Airlift Squadron, "After Action Report for MOSAIC TIGER," Memo to 317 OG/CC, December 27, 2021.

317th Airlift Wing, "After Action Report for Operation RAPID FORGE 2019," Memorandum for AMC/A3Y, August 12, 2019. Not available to the public.

ACC—*See* Air Combat Command.

Ackerman, Robert K., "Operation Enduring Freedom Redefines Warfare," *Signal Magazine*, September 2002.

Adam, John M., "Rethinking On-Call Airdrop for the C-130J: Concept Through Execution," U.S. Air Force Weapons School, research paper, 2016.

AFDN—*See* Air Force Doctrine Note.

AFDP—*See* Air Force Doctrine Publication.

AFI—*See* Air Force Instruction.

AFPAM—*See* Air Force Pamphlet.

Air Combat Command, "Lead Wing Concept of Operations," v2.0, undated draft.

Air Combat Command, "Air Combat Command Names Lead Wings," January 2022.

Allard, Kenneth, *Command, Control, and the Common Defense*, rev. ed., National Defense University, 1996.

Allerheiligen, Nathan A., "Keep on Trucking: An Entrepreneurial Approach to Intratheater Airlift," Air University, February 2013.

Alonso, Ricardo, Wouter Dessein, and Niko Matouschek, "When Does Coordination Require Centralization?" *American Economic Review*, Vol. 98, No. 1, 2008.

AFPAM—*See* Air Force Pamphlet.

Air Force Doctrine Note 1-21, *Agile Combat Employment*, Department of the Air Force, December 2021.

Air Force Doctrine Publication 3-36, *Air Mobility Operations*, Department of the Air Force, June 2019.

Air Force Instruction 38-101, *Manpower and Organization*, Department of the Air Force, August 2019.

Air Force Manual 10-207, *Command Posts*, Department of the Air Force, April 2018.

Air Force Pamphlet 10-219, Vol. 2, *Civil Engineer Contingency Response and Recovery Preparations*, Department of the Air Force, October 2018.

Air Mobility Command Instruction 11-208, *Mobility Air Force Management*, Department of the Air Force, February 2017.

Arrow, K. J., *The Limits of Organization*, W. W. Norton, 1974.

Brady-Lee, Patrick L., "Supporting the Air Mobility Needs of Geographic Combatant Commanders: An Evaluation Using the Principal-Agent Construct," Air University, School of Advanced Air and Space Studies, June 2014.

Brown, Jeffrey, S., "Divergent Paths: The Centralization of Airlift Command, Control, and Execution," Air University, School of Advanced Air and Space Studies, June 2005.

Carter, Ted E., Jr., "Theater Air Mobility: Historical Analysis, Doctrine, and Leadership," *Air Force Journal of Logistics*, Vol. 24, No. 3, April 2000.

Cohen, Rachel S., "New 'Lead Wing' Deployment Plan for Combat Aircraft Is Being Tested, Refined," *Air Force Times*, May 25, 2021.

Connor, Kathryn, Michael Vaseur, and Laura H. Baldwin, *Aligning Incentives in the Transportation Working Capital Fund*, Santa Monica, Calif: RAND Corporation, RR-2438-TRANSCOM, 2019. As of September 21, 2022: https://www.rand.org/pubs/research_reports/RR2438.html

DAFI—*See* Department of the Air Force Instruction.

Department of the Air Force Instruction 10-401, *Air Force Operations Planning and Execution,* Department of the Air Force, January 2021.

Department of Defense, *DoD Dictionary of Military and Associated Terms*, March 2017.

Devereaux, Richard T., "Theater Airlift Management and Control—Should We Turn Back the Clock to Be Ready for Tomorrow?" Air University Press, 1993.

Grandori, Anna, "An Organizational Assessment of Interfirm Coordination Modes," *Organizational Studies*, Vol. 18, No. 6, 1997.

Joint Publication 3-0, *Joint Operations*, Joint Staff, June 2022.

Joint Publication 3-17, *Air Mobility Operations*, Joint Staff, February 2019.

Joint Publication 4-0, *Joint Logistics*, Joint Staff, February 2019.

JP—*See* Joint Publication.

Kallberg, Jan, and Stephen S. Hamilton, "Resiliency by Retrograded Communications: The Revival of Shortwave as a Military Communication Channel," *Army Cyber Institute Journal*, November 2020.

Long, James E., "Adequacy of Airbase Opening Operations Doctrine," U.S. Army Command and General Staff College, 2006.

McLean, Brian, "Reshaping Centralized Control/Decentralized Execution for the Emerging Operating Environment," *Over the Horizon: Multidomain Operations and Strategies*, March 13, 2017. As of September 21, 2022: https://overthehorizonmdos.wpcomstaging.com/?s=McLean+%22reshaping+centralized%22

Milgrom, Paul, and John Roberts, *Economics, Organization, and Management*, Pearson, 1992.

Minihan, Mike, "AMC Mission Type Orders Primer," Air Mobility Command, March 2022.

Myers, Richard B., "A Word from the Chairman," *Joint Forces Quarterly*, No. 33, Winter 2002–2003.

Norwood, J. Scott, *Thunderbolts and Eggshells: Composite Air Operations During Desert Storm and Implications for USAF Doctrine and Force Structure*, Air University Press, September 1994.

Office of the Secretary of Defense, *Annual Report to Congress: Military and Security Developments Involving the People's Republic of China 2013*, 2013.

Oppelaar, Isaiah, "C4ISR Concerns for Air Mobility Command," *Over the Horizon: Multidomain Operations and Strategies*, May 2017. As of January 18, 2023: https://overthehorizonmdos.wpcomstaging.com/?s=C4ISR+Concerns+for+Air+Mobility+Command

Overy, R. J., *The Air War: 1939–1945*, Europa, 1980.

PACAF—*See* Pacific Air Forces.

Pacific Air Forces, *Agile Combat Employment (ACE): PACAF Annex to Department of the Air Force Adaptive Operations in Contested Environments*, June 2020.

Pacific Air Forces Instruction 10-2101, "Operations: Pacific Air Mobility Operations," Pacific Air Forces, April 12, 2019.

Percival, William D., "Integrating Joint Intratheater Airlift Command and Control with the Needs of the Modern Army," U.S. Army Command and General Staff College, 2008.

Pettyjohn, Stacie L., and Jennifer Kavanagh, *Access Granted: Political Challenges to the U.S. Overseas Military Presence, 1945–2014*, RAND Corporation, RR-1339-AF, 2016. As of September 21, 2022:
https://www.rand.org/pubs/research_reports/RR1339.html

Priebe, Miranda, Alan J. Vick, Jacob L. Heim, and Meagan L. Smith, *Distributed Operations in a Contested Environment: Implications for USAF Force Presentation*, RAND Corporation, RR-2959-AF, 2019. As of September 21, 2022:
https://www.rand.org/pubs/research_reports/RR2959.html

Pyles, Raymond A., Robert S. Tripp, Kristin F. Lynch, Don Snyder, Patrick Mills, and John G. Drew, *A Common Operating Picture for Air Force Materiel Sustainment*, RAND Corporation, MG-667-AF, 2008. As of September 21, 2022:
https://www.rand.org/pubs/monographs/MG667.html

Rifensburg, Gerard L., *Command and Control of the NATO Central Region Air Forces*, Air War College, Research Report, March 1989.

Rivkin, Jan W., and Nicolaj Siggelkow, "Balancing Search and Stability: Interdependencies Among Elements of Organizational Design," *Management Science*, Vol. 49, No. 3, March 2003.

Scharpf, Fitz W., "Games Real Actors Could Play: Positive and Negative Coordination in Embedded Negotiations," *Journal of Theoretical Politics*, Vol. 6, No. 1, 1994.

Slazinik, Ian M., *Air Mobility Future: Evolving Command and Control Relationships in the Information Age*, Air Force Institute of Technology, Graduate School of Engineering and Management, June 2016.

Snyder, Don, Bernard Fox, Kristin F. Lynch, Raymond E. Conley, John A. Ausink, Laura Werber, William Shelton, Sarah A. Nowak, Michael R. Thirtle, and Albert A. Robbert, *Assessment of the Air Force Materiel Command Reorganization: Report for Congress*, RAND Corporation, RR-389-AF, 2013. As of September 21, 2022:
https://www.rand.org/pubs/research_reports/RR389.html

Snyder, Don, Kristin F. Lynch, Colby Peyton Steiner, John G. Drew, Myron Hura, Miriam E. Marlier, and Theo Milonopoulos, *Command and Control of U.S. Air Force Combat Support in a High-End Fight*, RAND Corporation, RR-A316-1, 2021. As of September 21, 2022:
https://www.rand.org/pubs/research_reports/RRA316-1.html

Sutton, H. I., "How Russian Spy Submarines Can Interfere with Undersea Internet Cables," *Forbes*, August 2020.

Thompson, Jeffrey R., *Organizing for the Future: Aligning U.S. Air Force Cyber Support with Mission Assurance*, Air University, December 2011.

Tripp, Robert S., Kristin F. Lynch, Charles Robert Roll, Jr., John G. Drew, and Patrick Mills, *A Framework for Enhancing Airlift Planning and Execution Capabilities Within the Joint Expeditionary Movement Systems*, RAND Corporation, MG-377-AF, 2006. As of September 21, 2022: https://www.rand.org/pubs/monographs/MG377.html

Underwood, Kimberly, "Putting ACE to the Test: The Pacific Air Forces' Operation Pacific Iron 21 Goes In-Depth on Agile Combat Employment Capabilities," *Signal*, August 10, 2021.

Vick, Alan J., *Force Presentation in U.S. Air Force History and Airpower Narratives*, RAND Corporation, RR-2363-AF, 2018. As of September 21, 2022: https://www.rand.org/pubs/research_reports/RR2363.html

Waldon, Joseph P., "CHOP or Not? How Conventional C-130s Can Provide SOF with Effective Airlift," USAF Weapons School, research paper, 2015.

Wilson, Jeanne M., Susan G. Straus, and Bill McEvily, "All in Due Time: The Development of Trust in Computer-Mediated and Face-to-Face Teams," *Organizational Behavior and Human Decision Processes*, Vol. 99, No. 1, 2006.

Winnefeld, James A., and Dana J. Johnson, *Command and Control of Joint Air Operations: Some Lessons Learned from Four Case Studies of an Enduring Issue*, RAND Corporation, R-4045-RC, 1991. As of September 21, 2022: https://www.rand.org/pubs/reports/R4045.html